Lecture Notes of the Institute for Computer Sciences, Social-Informatics and Telecommunications Engineering 39

W0090974

Eitan Altman   Iacopo Carrera
Rachid El-Azouzi   Emma Hart   Yezekael Hayel (Eds.)

Bioinspired Models of Network, Information, and Computing Systems

4th International Conference, BIONETICS 2009
Avignon, France, December 9-11, 2009
Revised Selected Papers

 Springer

Volume Editors

Eitan Altman
INRIA, 2004 Route des Lucioles
06902 Sophia Antipolis Cedex, P.O. Box , France
E-mail: eitan.altman@sophia.inria.fr

Iacopo Carrera
CREATE-NET
vial alla Cascata 56D, 38123 Provo, Trento, Italy
E-mail: iacopo.carrera@create-net.org

Rachid El-Azouzi
Yezekael Hayel
LIA, University of Avignon
339 chemin des Meinajariès
84911 Avignon Cedex 9, France
E-mail: {rachid.elazouzi; yezekael.hayel@univ-avignon.fr

Emma Hart
Napier University, School of Computing
Merchiston Campus, 10 Colinton Road
Edinburgh, EH10 5DT, Scotland
E-mail: e.hart@napier.ac.uk

Library of Congress Control Number: 2010927025

CR Subject Classification (1998): C.2, H.4, D.2, C.2.4, D.1.3, D.4

ISSN	1867-8211
ISBN-10	3-642-12807-6 Springer Berlin Heidelberg New York
ISBN-13	978-3-642-12807-3 Springer Berlin Heidelberg New York

springer.com

© ICST Institute for Computer Sciences, Social-Informatics and Telecommunications Engineering 2010
Printed in Germany

Typesetting: Camera-ready by author, data conversion by Scientific Publishing Services, Chennai, India
Printed on acid-free paper   SPIN: 06/3180   5 4 3 2 1 0

Preface

This volume of LNICST is a collection of the papers of the 4th International Conference on Bio-Inspired Models of Network, Information, and Computing Systems (Bionetics). The event took place in the medieval city of Avignon, known also as the City of the Popes, during December 9 to 11, 2009. Bionetics main objective is to bring bio-inspired paradigms into computer engineereing and networking, and to enhance the fruitful interactions between these fields and biology.

The program of the conference indeed includes applications of various paradigms that have their origin in biology: population dynamics, branching processes, ant colony optimization. The proceedings include 19 papers covering a broad range of important issues in areas related to bio-inspired technologies. They correspond to presentations at 6 technical sessions. Four papers correspond to an invited session on the Epidemic-type forwarding in DTNs (sparse mobile ad-hoc wireless networks) organized by Dr Francesco De Pellegrini, (Italy, CREATE-NET). The following 9 papers (selected out of 15 submissions) correspond to contributions to regular sessions on Bio-inspired security, Bio-Inspired Networking, Bioinspired algorithms and software systems. The remaining 6 papers (selected out of a total of 9 submissions) are dedicated to work in progress. For each paper, we have provided at least two independent reviews, most of which were offered by members of the TPC.

Four keynote talks were presented at the conference by the following outstanding scientists: George KESIDIS (Professor of Electrical Engineering and Computer Science and Engineering, The Pennsylvania State University) who presented "Epidemiology of the spread of virus/worms in the Internet", Vivek S. BORKAR (Professor, School of Technology and Computer Science, Tata Institute of Fundamental Research, India) who gave a talk on "Variation of ant colony optimization for network problems", Wolfgang BANZHAF (Professor and Head Department of Computer Science, Memorial University of Newfoundland, Canada) who gave a talk on "Science and Engineering of Complex Systems". Finally, Nicolas CHAMPAGNAT (CR, INRIA Sophia Antipolis - Mediterranee) presenteed a talk on modelling Darwinian evolution, resulting from the interplay of phenotypic variation and natural selection through ecological interactions. We are very grateful for the participation of these speakers.

The conference included a session in the memory of Prof Thomas Vincent who was among the founders of evolutionary game theory. He was a professor in the Aerospace and Mecanical Engineering, and was a pioneer in bringing bio-inspired techniques into engineering. The session was composed of the keynote talk of Nicolas Champagnat as well as three invited talks by Pierre Bernhard, Bruno Gaujal and Eitan Altman. The success of this conference along with this proceeding volume is due in a large extent to the devoted work of the 28 TPC members to whom we thank. Special thanks are due to the conference steering committee and organizing committee for their help that made our job much easier and enjoyable. Warm thanks go to Ephie Deriche, Amar Azad, Tembine Hamidou, Tania Jimenez, Issam Mabrouki and Maria Morozova. We wish to thank our sponsors - ICST, Createnet, PerAda, INRIA, University of Avignon and Bionets.

Eitan Altman
Iacopo Carreras
Rachid El-Azouzi
Emma Hart
Yezekeal Hayel

Organization

Steering Committee

Imrich Chlamtac, Chair	Create-Net, Italy
Iacopo Carreras	Create-Net, Italy
Falko Dressler	Univ. of Erlangen, Germany
Tatsuya Suda	Univ. of California, Irvine and NTT DoCoMo, Inc., USA

Conference General Co-chairs

Eitan Altman	INRIA, France
Yezekael Hayel	University of Avignon, France

Technical Program Committee Co-chairs

Iacopo Carreras	Create-Net, Italy
Emma Hart	Edinburgh Napier University
Rachid El-Azouzi	Rachid El-Azouzi

Local Arrangement Co-chairs

Ephie Deriche	INRIA, France
Yezekael Hayel	University of Avignon, France
Hamidou Tembine	University of Avignon, France

Web Chair

Ephie Deriche	INRIA, France

Workshop Chair

Jian-Quin Liu	NICT, Japan
Tadashi Nakano	University of California, Irvine, USA

Publications Chair

Hamidou Tembine	University of Avignon, France

Publicity Chair

Amar Azad	INRIA, France
Issam Mabrouk	INRIA and University of Avignon, France

Conference Coordinator

Maria Morozova   ICST

Technical Program Committee

Jose Aguilar	University of Los Andes, Venezuela
Ozgur Akan	Middle East Technical University, Turkey
Ahmed Al-Hanbali	Eurandom, Netherlands
Peter Bentley	University College London United Kingdom
Falko Dressler	University of Erlangen, Germany
Francesco De Pellegrini	CREATE-NET, Italy
Niloy Ganguly	Indian Institute of Technology Kharagpur, India
Mark Jelasity	University of Szeged, Hungary
Anurag Kumar	Electrical Communication Engineering, Indian Institute of Science, Bangalore India
Kenji Leibnitz	Osaka University, Japan
Chris McEwan	Napier University, United Kingdom
Daniele Miorandi	CREATE-NET, Italy
Corrado Moiso	Telecom Italia Lab, Italy
Yuki Moritani	NTT DoCoMo Inc. Japan
Tadashi Nakano	University of California, Irvine United States
Ben Paechter	Napier University, United Kingdom
Marinella Petrocchi	IIT-CNR, Italy
Daniel Schreckling	University of Passau
Tatsuya Suda	University of California, Irvine United States
Jun Suzuki	Department of Computer Science, University of Massachusetts, Boston United States
Hamidou Tembine	LIA, University of Avignon, France
Gianluca Tempesti	University of York, United Kingdom
Antonio Manzalini	Telecom Italia Lab, Italy
Piet Van Mieghem	Delft University of Technology, Netherlands
Athanasios(Thanos) Vasilakos	University of Western Macedonia, Greece
Naoki Wakamiya	Osaka University, Japan
Lidia Yamamoto	University of Basel Switzerland
Lixia Zhang	UCLA, United States

Workshop Organisers

Marinella Petrocchi	IIT-CNR, Italy
Daniele Quercia	MIT, USA

Workshop Programme Committee

Alfarez Abdul-Rahman	Redkey Digital, UK
Eleonora Borgia	IIT-CNR, Italy
Neal Lathia	University College London, UK
Gabriele Lenzini	Novay, NL
Nadia Pisanti	University of Pisa, Italy
Daniel Schreckling	University of Passau, Germany
Lidia Yamamoto	University of Basel, CH
Christian Wallenta	Oxford University, UK

Table of Contents

Regular Session 3: Epidemic-Style Forwarding in DTNs

Regular Session 4: Bio-inspired Algorithms and Software Systems

Work in Progress Session 2

Cooperation in Hunting and Food-Sharing: A Two-Player Bio-inspired Trust Model

Ricardo Buettner

FOM Hochschule für Oekonomie & Management - University of Applied Sciences,
Arnulfstraße 30, 80335 Muenchen, Germany
ricardo.buettner@fom.de

Abstract. This paper proposed a new bilateral model supporting co-operative behavior. It is inspired by cooperation in hunting [34,38] and food sharing of female vampire bats [56,57,58]. In this paper, it is postulated, that low bounding of food capacity (fast saturation) in conjunction with a high demand of food energy (fast starving without food) strongly supports cooperative behavior. These postulations are integrated within the proposed model as an extension of the prisoner dilemma [10,11,49].

Keywords: bio-inspired models, trust management, self-organizing communities, cooperative systems, cooperative hunting, food sharing behavior, vampire bats.

1 Problem

From a collective perspective cooperative behavior is very important, but at first sight not of an individual perspective. The simple analysis of cooperative behavior in prisoner dilemma advises non-cooperation as the dominant strategy in the case of a lack of trust between the players [10,11,49]. But in nature many animals are not opportunistic, in fact they show cooperative behavior among each other in many cases. A lot of research took place to explain the differences between the advised dominant strategy from game theory compared to cooperative behavior in nature; e. g., iterated games [3], evolution-inspired games (kin selection [13,14], reciprocal altruism [27,33,48], master-and-servant strategy), sociological-inspired games (social identity theory [46,60]), or the possibility of punishment in case of non-cooperative behavior (folk theorem). But up to now, there is no satisfying explanation.

This is why, this paper focuses on interesting findings in biology concerning trust and cooperative behavior and found some inspirations of cooperative hunting behavior [34,38] as well as of food sharing behavior of vampire bats [56,57,58]. In this paper, it is postulated, that low bounding of food capacity (fast saturation) in conjunction with a high demand of food energy (fast starving without food) strongly supports cooperative behavior.

This paper is divided into 6 parts: After the problem description in section 1, a brief literature review concerning cooperation and competition in artificial intelligence is given in section 2. After that, some interesting aspects of cooperation

E. Altman et al. (Eds.): Bionetics 2009, LNICST 39, pp. 1–10, 2010.

and competition in nature are shown in section 3. On that biological-inspired basis, a new model as an extension of the basic iterated prisoner dilemma is proposed in section 4. In section 5 the proposed model is evaluated via simulation. Finally, in section 6 limitations of the work and future research directions are shown.

2   Cooperation and Competition in Artificial Intelligence

2.1   Game Theory

Despite of the fact, that trust and negotiations had already played an important role within the Babylonian Talmud [2], *G. Leibniz* [21] was one of the first who researched during the 17th century on concurrent and cooperative human behavior. *A. Cournot* [7] ascertained in the first half of the 18th century the main issues of negotiations in duopolies. Later, during the second half of the 18th century *F. Edgeworth* [9] and *J. Bertrand* [4] had proceeded the research, e. g., on graphical explorations, before *E. Zermelo* [61] provided the first mathematical formal approach on the basis of the Minimax-search in games in 1912. In 1921, *F. Borel* [6] introduced the concept of mixed strategies. On that basis, in 1928 *J. von Neumann* [53] proofed that every Two-Player-Game has a Minimax-Equilibrium in mixed strategies. Later, in 1944 *J. von Neumann* and *O. Morgenstern* [54] presented the influential work 'Theory of Games and Economic Behavior'. A further milestone was placed by *A. Tucker* [49] by the famous prisoner dilemma. *M. Dresher* [8] and *M. Flood* (e. g., [10,11,12]) from the *RAND Corporation* where the first who used systematically the prisoner dilemma in experiments. *R. Axelrod* [3] firstly implemented the prisoner dilemma in computer programs.

During the 1950ies and 1960ies most of the publications had focused on cooperative negotiation behavior. After that period, the research focus has moved to the non-cooperative branch. The most influential milestone in this research field was placed by *J. Nash* [30,31] with the later so-called 'Nash-Equilibrium'. *J. Nash* analyzed Two-person negotiation problems under the assumption of complete information. In 1960, *T. Schelling* [39] bridged game theory and general equilibrium conditions in an economy by introducing the 'focal point'. *W. Vickrey* [52] presented in 1961 the later so-called 'Vickrey-Auction Model' to identify true preferences of negotiation partners [52]. *R. Selten* [40,41,42] introduced the concept of 'Teilspielperfektheit' for sequential negotiations in 1965 and enormously stimulated business sciences with game-theoretical applications in the field of negotiations. Later, *D. Kreps* and *R. Wilson* [20] extended these works. A next important work was published by *J. Harsanyi* and *R. Selten* [16] by extending the work of *J. Nash* [29,30,31] to negotiation situations with incomplete information. One next step was the adaption of elements of the evolution theory into game theory. The corresponding concept of 'Evolutionary Stable Strategies' was introduced by *J. Smith* [47] in 1972. Finally, strategic behavior and interactions between self-interested agents were firstly analyzed by *J. Rosenschein* and *G. Zlotkin* [36,37,62] on the basis of the fundamental game-theoretic work [54,15,16,19].

2.2 Artifical Intelligence

The basis of Artificial Intelligence (AI) was generated by *W. McCulloch* and *W. Pitts* [26]. They proposed a biological-inspired artificial neural network based on the formal logic of *A. Whitehead* and *B. Russell* [55] and the Turing machine of *A. Turing* [50,51]. In 1956, *J. McCarthy* [25] introduced the name 'Artificial Intelligence' during a workshop in Dartmouth, New Hampshire, USA. *J. McCarthy* [24,23] defined AI as the science to design intelligent machines or rather intelligent programs.

Further major milestones in AI research had placed by *H. Simon* and *A. Newell* [32] with the 'General Problem Solver' and by *E. Shortliffe* [44,43] with the expert system 'MYCIN', before *M. Minsky* [28] postulated the thesis that 'intelligence' is generated by the interaction of a lot of simple modules. That was the key assumption to pave the way for 'Distributed Artificial Intelligence' (DAI).

2.3 Software Agents as Biological-inspired Programs

Software agents were developed as a part of DAI research. Within this research area, 'Distributed Problem Solving' (DPS) can be separated from 'Multi-Agent-Systems' (MAS). The concept of a software agent is based on the actor model by *C. Hewitt* [18]. The local node within a DPS system is not independent from the system [5]. In contrast, in MAS a software agent is independent and decides its participation by its own [36]. Key biological-inspired characteristics of software agents are autonomy, social ability, reactivity and pro-activeness [59]. Because of this characteristics, in an MAS is no supervisor who controls the software agents, particularly punish non-cooperative behavior. Other trust-supported mechanisms are needed. Here, analogies can be found within animality. To support trust in cooperative behavior, some animals show solutions, especially in hunting scenes and food sharing.

3 Aspects of Cooperation and Competition in Nature

In nature many animals show cooperative behavior; e. g., kin selection [13,14], reciprocal altruism [27,33,48], cooperative hunting [34,38], or food sharing [56,57,58]. In the following it is focused on cooperative hunting behavior and on food sharing behavior of vampire bats.

3.1 Cooperative Hunting

C. Packer and *L. Rutton* [34] reviewed the cooperative hunting literature and analyzed the advantages and problems of cooperative hunting. In case of a larger prey, cooperative hunting is often the observed strategy in nature. *D. Scheel* and *C. Packer* [38] generally pointed out that cooperative behavior in hunting depends on the size of prey.

However, *C. Packer* and *L. Rutton* [34] reviewed data from 28 studies of group hunting and showed that hunting success generally increases asymptotically with increasing group size in circumstances where individuals are expected to hunt cooperatively. In small groups every individual is needed and have to participate to be successful in hunting.

3.2   Food Sharing Behavior of Female Vampire Bats

G. Wilkinson [56,57,58] have extensively researched on food-sharing behavior of female vampire bats (lat. 'desmodus rotundus'). He showed that food sharing by regurgitation of blood among wild vampire bats depends equally and independently on degree of relatedness. Vampire bats fail to secure a meal in approximately 10 percent of their foraging bouts, while approximately 33 percent of bats under two years of age fail [57]. Missed meals can have enormously effects on survival of the bats, because young vampire bats will starve to death after around 60 hours without a meal [58]. *G. Wilkinson* [56] found that reciprocal exchanges of blood meals by regurgitation are common between female vampires.

Female vampire bats live around 18 years. Group composition appears to be stable over long time [56,57]. This is why multiple interactions between the bats are most likely. Laboratory and field studies indicate that bats are significantly less likely to provide a meal to those bats who failed to reciprocate this action in the past [27, p. 275].

In summary, three major keys seems to support cooperative behavior:

(k_a) small groups, and
(k_b) a high demand of food energy (fast starving without food), and
(k_c) low bounding of food capacity (fast saturation).

4   Formal Model

4.1   Initial Assumptions

A1. There are two agents, A and B (see k_a).
A2. Each agent (A and B) acts rational within market economy conditions.
A3. Each agent (A and B) wants to maximize its own utility ($u_{A,B}$).
A4. Neither, A nor B has any information about the opponent.
A5. The game is played repeatedly (iterated game with $1..i..N$ rounds).

4.2   Basic Model

The basic model correspondents with the prisoner dilemma [10,11,49].

Definition 1. *At each round i, A and B can choose privately one of the following possible strategies $S(i)_{A,B} = [C, D]$:*

Cooperation: *Here, the agent wants to cooperate and tries to share the cake fifty-fifty.*
Deception: *Here, the agent tries to cheat.*

Definition 2. *Depending on the chosen strategy $S_{A,B} = [C, D]$ the following symmetric payoff function $[u_A|u_B]$ exists $(T > R > P > S)$:*

$$[u_A|u_B] = f(S_A, S_B) = \begin{array}{c} [S_B] \\ {}_{[S_A]} [u_A|u_B] \end{array} = \begin{bmatrix} C \\ D \end{bmatrix} \begin{bmatrix} C & D \\ R|R & S|T \\ T|S & P|P \end{bmatrix} \tag{1}$$

4.3   Model Extensions

E1. Each agents survive itself in the case of positiv energy $(e(i)_{A,B} \geq 0)$.
E2. There is only a cake to divide or rather a payoff in the case of both agents A and B are alive $(e(i)_A \geq 0$ AND $e(i)_B \geq 0)$. (For successful hunting in small groups every agent is needed and has to participate (see k_a).)
E3. At each round i, A and B have to spend s fix energy points (see k_b).
E4. The energy capacities of A and B are bounded to e_A^{max}, e_B^{max}. More energy payoffs from the payoff function $[u_A|u_B]$ runs to seed (see k_c).

5   Evaluation

According to [17], in order to show the utility, quality, and efficacy of the proposed model as a design artifact, it has to be evaluated via well-executed evaluation methods; e.g. by simulation. During a simulation the model will be checked with artificial data. Goal of the evaluation is to show the benefits of the proposed model compared to other models.

5.1   Simulation Setting

Five extended meta-strategies are utilized for simulation: 1. Strong Cooperation (COOP), 2. Strong Deception (DEC), 3. Random Cooperation or Deception (RAND), 4. Tit for Tat (TFT) [3], and 5. Tit for two Tats (TFTT).

5.2   Experimental Results

The simulation results of non-redundant combinations of the meta-strategies (COOP, DEC, RAND, TFT, TFTT) of A and B are presented in Tab. 2.

Table 1. Simulation Parameters

Parameter	Value(s)
Starting energy:	$e(0)_{A,B} = 6$
Fixed consumption energy:	$s = 3$
Energy capacity:	$e_{A,B}^{max} = 10$
Payoff matrix:	$T = 10, R = 5, P = 2, S = 0$

Table 2. Results

Strategy A	Strategy B	Rounds (i) Survive A	Rounds (i) Survive B
COOP	COOP	∞	∞
COOP	DEC	2	5
COOP	RAND	$ND(\mu = 9.80; \delta = 8.09)$	$ND(\mu = 13.20; \delta = 8.18)$
COOP	TFT	∞	∞
COOP	TFTT	∞	∞
DEC	DEC	6	6
DEC	RAND	$ND(\mu = 6.20; \delta = 0.76)$	$ND(\mu = 3.10; \delta = 0.92)$
DEC	TFT	6	4
DEC	TFTT	5	2
RAND	RAND	$ND(\mu = 21.87; \delta = 20.07)$	$ND(\mu = 21.60; \delta = 19.74)$
RAND	TFT	$ND(\mu = 365.57; \delta = 349.49)$	$ND(\mu = 364.53; \delta = 349.73)$
RAND	TFTT	$ND(\mu = 52.37; \delta = 66.34)$	$ND(\mu = 49.77; \delta = 66.66)$
TFT	TFT	∞	∞
TFT	TFTT	∞	∞
TFTT	TFTT	∞	∞

5.3 Discussion

As shown in Tab. 2 the extended model intensively supports cooperation. As long as no agent tries to cheat, both agents survive unlimited time. The special variants (TFT) and (TFTT) are leap of faith meta-strategies while (COOP) is the strong cooperation meta-strategy. When both agents use one of these meta-strategies they survive endlessly.

On the other hand, in all variants, deception is strongly punished. If one of the agents use (DEC) or (RAND) both agents die.

6 Conclusion

The proposed model strongly supports cooperation of agents (Tab. 2). Because of bounding of the energy capacities of both agents in combination with a high demand of energy at every round, agents in small groups are forced to be cooperative. These assumptions are inspired from nature and concern the trust in systems (general conditions, see *N. Luhmann* [22]), not the trust in other agents. Non-cooperation quickly ends in starving of both agents.

There are practical economic implications: Because of the possibility that agents can hoard money and goods, non-cooperative behavior is emphasized. To support cooperative behavior between agents in value-added chains, continuous low-level deprecation of money is an appropriate instrument, e. g., by a moderate inflation rate and a progressive wealth tax.

6.1 Limitations

The proposed model is limited to two agents (see assumption A1). Further, rational behavior of the agents is assumed. But, since [45] it is clear that real agents act bounded rational. Despite of a robustness check by variation of the parameters in table 1 the model was only checked with some artificial data. An intensive evaluation or a mathematical proof of the model would be helpful.

6.2 Further Research Directions

A first extension of the proposed model should be the relaxation to more than two agents. Furthermore, other meta-strategies (e. g., mixed-strategies, customized) should be considered within the evaluation. Finally, the suggested implications to economy (inflation rate and progressive wealth taxes) should be economically and politically checked.

References

1. Proceedings of the Fifth International Congress of Mathematicians, Cambridge, UK, August 22-28, 1912, vol. 2. Cambridge University Press, Cambridge (1913)
2. Aumann, R.J., Maschler, M.: Game Theoretic Analysis of a Bankruptcy Problem from the Talmud. Journal of Economic Theory 36(2), 195–213 (1985)
3. Axelrod, R.M.: The Evolution of Cooperation. Basic Books, New York (1984)
4. Bertrand, J.L.F.: Théorie Mathématique de la Richesse Sociale. Journal des Savants 67, 499–508 (1883)
5. Bond, A.H., Gasser, L.: An Analysis of Problems and Research in DAI. In: Bond, A.H., Gasser, L. (eds.) Readings in Distributed Artificial Intelligence, pp. 3–35. Morgan Kaufmann Publishers, San Mateo (1988)
6. Borel, F.E.J.E.: La théorie du jeux et les équations intégrales à noyau symétrique. Comptes rendus hebdomadaires des séances de l'Académie des sciences 173, 1304–1308 (1921); translated as 'The Theory of Play and Integral Equations with Skew Symmetric Kernels' by Leonard J. Savage. Econometrica 21(1), 97–100 (1953)
7. Cournot, A.A.: Recherches sur les Principes Mathématiques de la Théorie des Richesses. L. Hachette et Cie., Paris (1838)
8. Dresher, M.: Games of Strategy: Theory and Applications. Prentice-Hall, Englewood Cliffs (1961)
9. Edgeworth, F.Y.: Mathematical Psychics: An Essay on the Application of Mathematics to the Moral Sciences. Kegan Paul, London (1881); Reprinted New York: Augustus M. Kelley (1967)

10. M. M. Flood. A Preference Experiment. Report, RAND Corperation(1951)
11. M. M. Flood. A Preference Experiment (Series 2, Trial 1). Report, RAND Corperation (1951)
12. M. M. Flood. A Preference Experiment (Series 2, Trials 2, 3, 4). Report, RAND Corperation (1952)
13. Hamilton, W.D.: The genetical evolution of social behaviour, I. Journal of Theoretical Biology 7(1), 1–16 (1964)
14. Hamilton, W.D.: The genetical evolution of social behaviour, II. Journal of Theoretical Biology 7(1), 17–52 (1964)
15. Harsanyi, J.C.: Games with Incomplete Information Played by 'Bayesian' Players, I-III, part I: The Basic Model. Management Science 14(3), 159–182 (1967)
16. Harsanyi, J.C., Selten, R.: A Generalized Nash Solution for Two-Person Bargaining Games with Incomplete Information. Management Science 18(5), P80–P106 (1972)
17. Hevner, A.R., March, S.T., Park, J., Ram, S.: Design Science in Information Systems Research. MIS Quarterly 28(1), 75–105 (2004)
18. Hewitt, C.: Viewing Control Structures as Patterns of Passing Messages. Artificial Intelligence 8(3), 323–364 (1977)
19. Kreps, D.M., Milgrom, P.R., Roberts, J., Wilson, R.B.: Rational Cooperation in the Finitely Repeated Prisoners' Dilemma. Journal of Economic Theory 27(2), 245–252 (1982)
20. Kreps, D.M., Wilson, R.B.: Sequential equilibria. Econometrica 50(4), 863–894 (1982)
21. Leibniz, G.W.: Nouveaux essais sur l'entendement humain. 1704. Completed 1704, Published 1765, Dt. Übersetzung und Einleitung von Ernst Cassirer: Neue Abhandlungen über den menschlichen Verstand. Meisner Verlag (1915)
22. Luhmann, N.: Vertrauen: Ein Mechanismus der Reduktion von Komplexität. Enke Verlag, Stuttgart (1968)
23. McCarthy, J.: Formalization of Common Sense, Papers by John McCarthy. Ablex, Norwood (1990)
24. McCarthy, J., Hayes, P.J.: Some Philosophical Problems from the Standpoint of Artificial Intelligence. In: Meltzer, B., Michie, D. (eds.) Machine Intelligence 4, pp. 463–502. Edinburgh University Press (1969)
25. McCarthy, J., Minsky, M., Rochester, N., Shannon, C.E.: A Proposal for the Dartmouth Summer Research Project on Artificial Intelligence. Projektantrag (August 1955), http://www-formal.stanford.edu/jmc/history/dartmouth/dartmouth.html (08.09.2007)
26. McCulloch, W.S., Pitts, W.H.: A Logical Calculus of the Ideas Immanent in Nervous Activity. Bulletin of Mathematical Biophysics 5, 115–133 (1943)
27. Mesterton-Gibbons, M., Dugatkin, L.A.: Cooperation Among Unrelated Individuals: Evolutionary Factors. The Quarterly Review of Biology 67(3), 267–281 (1992)
28. Minsky, M.: The Society of Mind. Simon and Schuster, New York (1986)
29. Nash, J.F.: The Bargaining Problem. Econometrica 18(2), 155–162 (1950)
30. Nash, J.F.: Equilibrium Points in n-Person Games. Proceedings of the National Academy of Sciences 36, 48–49 (1950)
31. Nash, J.F.: Non-Cooperative Games. Annals of Mathematics 54(2), 286–295 (1951)
32. Newell, A., Simon, H.A.: A Program that Simulates Human Thought. In: Computers and Thought, pp. 279–293. McGraw-Hill, New York (1963)
33. Packer, C.: Reciprocal altruism in *Papio anubis*. Nature 265(5593), 441–443 (1977)

34. Packer, C., Rutton, L.: The Evolution of Cooperative Hunting. American Naturalist 132(2), 159–198 (1988)
35. Rasmusen, E.B. (ed.): Readings in Games and Information, 1st edn. Blackwell Readings for Contemporary Economics. Blackwell Publishers, Malden (2001)
36. Rosenschein, J.S.: Rational Interaction: Cooperation Among Intelligent Agents. PhD thesis, Computer Science Department, Stanford University, Stanford, California, USA (March 1985)
37. Rosenschein, J.S., Zlotkin, G.: Rules of Encounter: Designing Conventions for Automated Negotiation among Computers. MIT Press, Boston (1994)
38. Scheel, D., Packer, C.: Group hunting behaviour of lions: a search for cooperation. Animal Behaviour 41(4), 697–709 (1991)
39. Schelling, T.C.: The Strategy of Conflict. Harvard University Press, Cambridge (1960)
40. Selten, R.: Spieltheoretische Behandlung eines Oligopolmodells mit Nachfrageträgheit. Teil 1: Bestimmung des dynamischen Preisgleichgewichts. Zeitschrift für die gesamte Staatswissenschaft (ZgS) 121(2), 301–324 (1965)
41. Selten, R.: Spieltheoretische Behandlung eines Oligopolmodells mit Nachfrageträgheit. Teil 2: Eigenschaften des dynamischen Preisgleichgewichts. Zeitschrift für die gesamte Staatswissenschaft (ZgS) 121(4), 667–689 (1965)
42. Selten, R.: Reexamination of the Perfectness Concept for Equilibrium Points in Extensive Games. International Journal of Game Theory 4(1), 25–55 (1975)
43. Shortliffe, E.H.: MYCIN: Computer-Based Medical Consultations. American Elsevier, New York, NY, USA (1976). Based on a PhD thesis, Stanford University, Stanford, CA, USA (1974)
44. Shortliffe, E.H., Buchanan, B.G.: A Model for Inexact Reasoning in Medicine. Mathematical Biosciences 23(3-4), 351–379 (1975)
45. Simon, H.A.: Administrative Behavior: A Study of Decision-Making Processes in Administrative Organizations, 1st edn. Free Press, New York (1947); orig. 1945 Chicago / 2nd edn. 1965/ 3rd edn. 1976 / 4th edn. 1997
46. Simpson, B.: Social Identity and Cooperation in Social Dilemmas. Rationality and Society 18(4), 443–470 (2006)
47. Smith, J.M.: Game Theory and the Evolution of Fighting. In: On Evolution, pp. 8–28. Edinburgh University Press, Edinburgh (1972)
48. Trivers, R.L.: The Evolution of Reciprocal Altruism. The Quarterly Review of Biology 46(1), 35–57 (1971)
49. Tucker, A.W.: A Two-Person Dilemma. Stanford University mimeo., unpublished. Reprinted in [35, S. 7f.] (May 1950)
50. Turing, A.M.: On Computable Numbers, with an Application to the Entscheidungsproblem. Proceedings of the London Mathematical Society s2-42(1), 230–265 (1937)
51. Turing, A.M.: On Computable Numbers, with an Application to the Entscheidungsproblem. A Correction. Proceedings of the London Mathematical Society s2-43(6), 544–546 (1938)
52. Vickrey, W.: Counterspeculation, Auctions, and Competitive Sealed Tenders. Journal of Finance 16(1), 8–37 (1961)
53. von Neumann, J.L.: Zur Theorie der Gesellschaftsspiele. Mathematische Annalen 100(1), 295–320 (1928)
54. von Neumann, J.L., Morgenstern, O.: Theory of Games and Economic Behavior, 1st edn. Princeton University Press, Princeton (1944); 2 edn. 1947, 3 edn. 1953
55. Whitehead, A.N., Russell, B.A.W.: Principia Mathematica, 1st edn., vol. 1-3. Cambridge University Press, Cambridge (1910, 1912, 1913)

56. Wilkinson, G.S.: Reciprocal food sharing in the vampire bat. Nature 308(5964), 181–184 (1984)
57. Wilkinson, G.S.: Reciprocal Altruism in Bats and Other Mammals. Ethology and Sociobiology 9(2-4), 85–100 (1988)
58. Wilkinson, G.S.: Food Sharing in Vampire Bats. Scientific American 262(2), 76–82 (1990)
59. Wooldridge, M., Jennings, N.R.: Intelligent Agents: Theory and Practice. Knowledge Engineering Review (KER) 10(2), 115–152 (1995)
60. Yamagishi, T., Kiyonari, T.: The Group as the Container of Generalized Reciprocity. Social Psychology Quarterly 63(2), 116–132 (2000)
61. Zermelo, E.: Über eine Anwendung der Mengenlehre auf die Theorie des Schachspiels. In: Proceedings of the Fifth International Congress of Mathematicians, [1], Cambridge, UK, August 22-28, pp. 501–504 (1912)
62. Zlotkin, G., Rosenschein, J.S.: Negotiation and Task Sharing Among Autonomous Agents in Cooperative Domains. In: IJCAI 1989: Proceedings of the Eleventh International Joint Conference on Artificial Intelligence, Detroit, MI, USA, August 20-25, pp. 912–917 (1989)

Executable Specification of Cryptofraglets in Maude for Security Verification

Fabio Martinelli and Marinella Petrocchi

IIT-CNR, Pisa, Italy

Abstract. Fraglets are computation fragments flowing through a computer network. They implement a chemical reaction model where computations are carried out by having fraglets react with each other. The strong connection between their way of transforming and reacting and some formal rewriting system makes a fraglet program amenable to verification. Starting from a threat model which we intend to use for modeling secure communication protocols with fraglets, we propose an executable specification of fraglets (and fraglets-based cryptographic protocols) in the rewriting logic-based Maude interpreter.

Keywords: Fraglets, cryptofraglets, threat model, security protocols, Maude.

1 Introduction

Fraglets [22,23,24,25] represent an execution model for communication protocols that resembles the chemical reactions in living organisms. It was originally proposed for making automatic the whole process of protocol development, involving the various phases of design, implementation and deployment. Fields of applications have been protocol resilience and genetic programming experiments.

The fraglets model has been adopted in the BIONETS EU project [2] and some security and trust extensions to the original model have become necessary to make it a running framework. Thus, in the past few years, fraglets have been extended i) with instructions for symmetric cryptography [21]; ii) with access control mechanisms [14]; iii) with dedicated primitives for trust management [15]. This work has been mainly done with the intent to use fraglets for modeling security protocols and verifying security properties within BIONETS. With an eye on verification, an encoding from the programming language defined for fraglets and the MultiSet Rewriting formal rules [3] has been shown in [21], as a first brick to give a formal semantics to fraglets, the starting point for developing verification tools.

It was indeed the similarity between the fraglets programming language and the rewriting systems, combined with the need of modeling and verifying security protocols for the bio-inspired networks BIONETS, that lead us to consider the executable specification language Maude [5,16], based on rewriting logic [18]. Maude offers useful advantages for formalizing and verifying security communication protocols (see, e.g., [1,6]. First, its efficient executability allows both prototyping and debugging of protocols specifications. Furthermore, since a concurrent system can have many different behaviours, exploring a single execution could not be sufficient to prove security

E. Altman et al. (Eds.): Bionetics 2009, LNICST 39, pp. 11–23, 2010.

properties of the system. Maude supports ad hoc defined strategies for exploring all the execution of a system.

The original contributions of the paper are the following. First, we define a threat model for fraglets by adding an adversary to the fraglet model of a secure communication network. Second, we present our executable specification of fraglets in Maude. Third, we show how to specify and execute a fraglets-based instantiation of the Needham-Schroeder Public Key protocol (NSPK). A security verification with respect to message secrecy is carried out by means of Maude built-in commands. The overall goal is the achievement of a general fraglets framework for the modeling and verification of security issues in communication protocols.

The paper is organized as follows. Section 2 recalls the fraglets model and introduces a refined version of cryptofraglets. Section 3 defines a threat model for cryptofraglets, by discussing the capabilities of an adversary that is going to subvert the standard operations in a secure fraglet network. We introduce the notion of the adversary's knowledge and we define the secrecy property for fraglets. Section 4 presents the executable specification of cryptofraglets in Maude, according to the defined threat model. Section 5 shows a fraglets-based instantiation of two interleaved sessions of NSPK. Finally, Section 6 gives some final remarks and discusses current limitations and future goals.

2   Fraglets

A fraglet is denoted as $[s1\ s2\ \ldots tail]$, where si $(1 \leq i \leq n)$ is a symbol and $tail$ is a (possibly empty) sequence of symbols. Nodes of a communication network may process fraglets as follows. Each node maintains a fraglet store to which incoming fraglets are added. Fraglets may be processed only within a store. The $send$ operation transfers a fraglet from a source store to a destination store.

Fraglets are processed through a simple prefix programming language. Transformation instructions involve a single fraglet, while reactions involve two fraglets. Table 1 shows the fraglets core instructions. The interested reader can find the comprehensive tutorial on [10].

Table 1. The set of fraglets core instructions

match	[match t tail1], [t tail2] $\rightarrow$ [tail1 tail2]
matchp	[matchp t tail1], [t tail2] $\rightarrow$ [matchp t tail1], [tail1 tail2]
send	s_A[send B tail] $\rightarrow$ s_B[tail]
nop	[nop tail] $\rightarrow$ [tail]
nul	[nul tail] $\rightarrow$ []
dup	[dup t a tail] $\rightarrow$ [t a a tail]
exch	[exch t a b tail] $\rightarrow$ [t b a tail]
fork	[fork a b tail] $\rightarrow$ [a tail], [b tail]
pop2	[pop2 h t a b tail] $\rightarrow$ [h a], [t b tail]
split	[split seq1 * seq2] $\rightarrow$ [seq1], [seq2]

Two fraglets react by instruction *match*, and their tails are concatenated. With the catalytic *matchp*, the reaction rule persists. Instruction *send* performs a communication between fraglets stores. It transfers a fraglet from store S_A to store S_B. Notation $S_A[s1\ s2\ \dots\ tail]$ denotes that the fraglet is located at S_A. The name of the destination store is given by the second symbol in the original fraglet $[send\ S_B\ tail]$. Where not strictly necessary, we omit to make the name of the store explicit.

Instruction *nop* does nothing, except consuming the instruction tag. Instruction *nul* destroys a fraglet. Finally, there are a set of transformation rules that perform symbol manipulation, like duplicating a symbol (*dup*), swapping two tags (*exch*), copying the tail and prepending different header symbols (*fork*), popping the head element a out of a list $a\ b\ tail$ (*pop2*), and finally breaking a fraglet into two at the first occurrence of symbol $*$ (*split*).

In [21,14], we proposed the cryptofraglets, which extend the fraglets programming language with basic cryptographic instructions. In defining cryptofraglets, we abstract from the cryptographic details concerning the operations by which they can be encrypted, decrypted, hashed, *etc.*. We make the so called *perfect cryptography assumption* and we consider encryption as a black box: an encrypted symbol, or sequence of symbols, cannot be correctly learnt unless with the right decryption key. This approach is standard in (most of) the analysis of cryptographic communication protocols, see, *e.g.*, [4,9,11,13].

Here, we give a slightly modified version of the crypto-instructions, originally seen as reaction rules, and here rephrased to transformation rules, for a more convenient specification in Maude (see Section 4).

Table 2. Crypto-instructions for encryption, decryption, and hashing

enc	$[enc\ newtag\ k_1\ tail] \rightarrow [newtag\ tail_{k_1}]$
dec	$[dec\ newtag\ k_2\ tail_{k_1}] \rightarrow [newtag\ tail]$
hash	$[hash\ newtag\ tail] \rightarrow [newtag\ h(tail)]$

The encryption instruction takes as input the fraglet $[enc\ newtag\ k_1\ tail]$, consisting of the reserved instruction tag *enc*, an auxiliary tag *newtag*, the encryption key k_1, and a generic sequence of symbols *tail*, representing the meaningful payload to be encrypted. It returns the fraglet $[newtag\ tail_{k_1}]$, with the auxiliary tag and the cyphertext $tail_{k_1}$. Decryption and hashing rules can be similarly explained. Note that, since fraglets processing is through matching tags, the presence of either a reserved instruction tag or an auxiliary tag as the leftmost symbol is necessary for the computation to go on.

We set $k_1 = k_2 = k$ when dealing with shared-key cryptography, and we consider a pair of public/private keys (pk, sk) when dealing with asymmetric cryptography, with, *e.g.*, $k_1 = sk$ and $k_2 = pk$. We employ hash functions to implement digital signatures, by applying the *enc* instruction with private key sk to the hash $h(tail)$. Signature verification is done by applying instruction *dec* with public key pk to the encrypted hash $h(tail)_{sk}$. However, in the rest of the paper, we will not consider digital signatures.

It is worth noticing that the set of fraglets programming instructions, given in Tables 1 and 2, consists of *rewrite rules* [16,18], with a simple *rewriting semantics* in which the left-hand side pattern (to the left of $\rightarrow$) is replaced by corresponding instances of the right-hand side. They represent *local transition rules* in a possibly distributed, concurrent system. Thus, we assume the presence of a *rewrite system* (defined by a single step transition operator $\rightarrow$, with $\rightarrow^*$ as its transitive and reflexive closure) operating on fraglets by means of the rewrite rules corresponding to the fraglets programming instructions. If we let f, f' range over fraglets, by applying operations from the rewrite system to a set F of fraglets, a new set $D(F) = \{ f \mid F \rightarrow^* f\}$ of fraglets can be obtained. As an example of a simple step transition rule application, we have: $D(\{[dup\ t\ a\ tail]\}) = \{ [dup\ t\ a\ tail], [t\ a\ a\ tail]\}$ since $[dup\ t\ a\ tail] \rightarrow_{dup} [t\ a\ a\ tail]\}$.

Below, we show the initial pool of fraglets, originally at stores $\mathcal{S}_A$ and $\mathcal{S}_B$, needed to execute a simple program that symmetrically encrypts a fraglet at store A, transfers the cyphertext at store B, and decrypts it at store B.

> *pool of fraglets originally at $\mathcal{S}_A$:*
> $_A[$KEY K$]$ $_A[$MSG M$]$
> $_A[$MATCH KEY MATCH MSG ENC NEWTAG$]$ $_A[$MATCH NEWTAG SEND B KMSG$]$

> *pool of fraglets originally at $\mathcal{S}_B$:*
> $_B[$KEY K$]$ $_B[$MATCH KEY MATCH KMSG DEC NEWTAG$]$

One possible execution of the program is as follows.

> $_A[$KEY K$]$ $_A[$MATCH KEY MATCH MSG ENC NEWTAG K$]$ $\rightarrow_{match}$ $_A[$MATCH MSG ENC NEWTAG$]$
> $_A[$MATCH MSG ENC NEWTAG$]$ $_A[$MSG M $]$ $\rightarrow_{match}$ $_A[$ENC NEWTAG M$]$
> $_A[$ENC NEWTAG M$]$ $\rightarrow_{enc}$ $_A[$NEWTAG $m_k]$
> $_A[$MATCH NEWTAG SEND B KMSG$]$ $_A[$NEWTAG $m_k]$ $\rightarrow_{match}$ $_A[$SEND B KMSG $m_k]$
> $_A[$SEND B KMSG $m_k]$ $\rightarrow_{send}$ $_B[$KMSG $m_k]$
> $_B[$KEY K$]$ $_B[$MATCH KEY MATCH KMSG DEC NEWTAG$]$ $\rightarrow_{match}$ $_B[$MATCH KMSG DEC NEWTAG K$]$
> $_B[$MATCH KMSG DEC NEWTAG K$]$ $_B[$KMSG $m_k]$ $\rightarrow_{match}$ $_B[$DEC NEWTAG K $m_k]$
> $_B[$DEC NEWTAG K $m_k]$ $\rightarrow_{dec}$ $_B[$NEWTAG M$]$

Tags *key*, *msg*, and *kmsg* are auxiliary. It is understood that $\mathcal{S}_A$ and $\mathcal{S}_B$ are the only stores at stake, and that, originally, there are no other fraglets than the ones in the initial pool.

3   A Threat Model for Fraglets

In this section we present a threat model for fraglets. We identify nodes of a communication network *A B C etc..* with their fraglets stores, *viz.* $\mathcal{S}_A$, $\mathcal{S}_B$, $\mathcal{S}_C$, etc.. Thus, principals of a communication protocol are fraglets stores, within which fraglets (protocol code + protocol messages) are processed. In particular, communications are via the *send* instruction.

We consider a protocol specification involving two honest roles, *viz.* an *initiator* $\mathcal{S}_S$ and a *responder* $\mathcal{S}_R$. Rather than a direct communication between them, we assume all their communication to flow through an *untrusted* store $\mathcal{S}_X$ which can either listen to or modify (fake) the fraglets exchanged between $\mathcal{S}_S$ and $\mathcal{S}_R$. Indeed, when modeling and verifying security properties of communication protocols, it is quite common to include an additional intruder (*à la* Dolev-Yao [7]) that is supposed to be malicious and whose aim is to subvert the protocol's correct behaviour. A protocol specification is then considered secure w.r.t. a security property if it satisfies this property despite the presence of the intruder. We thus propose a framework of three types of fraglets stores:

1. S_S plays the role of the protocol's initiator;
2. S_R plays the role of the protocol's responder;
3. S_X plays the role of the active and malicious intruder.

We let the initiator and the responder to be forced to communicate with the untrusted store through disjoint sets of communication (*send*) actions Σ_{com}^{S} and Σ_{com}^{R}, resp., such that a direct communication between them is impossible.

It is also assumed that, at deployment, each store S_I, with $I = \{S, R, X\}$, contains the pool of keys Λ^I needed for the store to perform encryptions and decryptions.

In Fig. 1 we have sketched the communication scenario described above and we have instantiated it with $\Sigma_{com}^{S} = \{s_S[send\ S_X\ ktail\ tail_K],$
$s_X[send\ S_S\ ktail\ tail_K]\}$ and $\Sigma_{com}^{R} = \{s_X[send\ S_R\ ktail\ tail_K],$
$s_R[send\ S_X\ ktail\ tail_K]\}$, for some generic auxiliary tag *ktail* and some generic encrypted sequence of symbols $tail_K$. We do not explicitly specify the type of the key used for encryption (and decryption), but we let $K \in \Lambda^I$. For instance, $\Lambda^S = \{k_{SR}, k_{SX}, sk_S, pk_R, pk_X\}$, *i.e.*, the shared secret key between S_S and S_R, the shared secret key between S_S and S_X, the secret key of S_S, and the public keys of S_X and S_R.

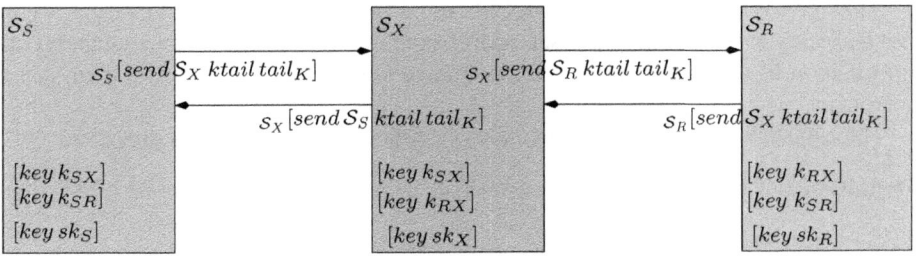

Fig. 1. The threat model for fraglets and fraglets stores

We do not fix *a priori* any specific behaviour for the intruder. S_X can process all the fraglets that it contains, by means of all the usual fraglets instructions. S_X can also honestly engage in a security protocol. Thus, it is decorated with its own pair of public/private keys (pk_X, sk_X), the symmetric key k_{SX}, shared with S_S, and the symmetric key k_{RX}, shared with S_R.

We also assume that i) private keys are initially contained only by the legitimate stores; ii) shared secret keys are initially contained only by the legitimate stores that share those keys. Finally, we assume that all the public keys are contained by all the stores at stake (public keys are not shown in the figure).

The intruder's knowledge. Here, we beat about the bush of the classical notion of the intruder's knowledge [8,20], *e.g.*, the set of all the messages the intruder knows from the beginning (its initial knowledge) united with the messages it can derive from the ones intercepted during a run of the protocol. Within a fraglet framework, this standard concept is slightly modified.

First, we say that a symbol is public the symbol is the second leftmost symbol of a fraglet at $\mathcal{S}_X$. Intuitively, the intruder's knowledge, at a given state of the computation, is the set of all the symbols that $\mathcal{S}_X$ knows. Let $\mathsf{F}_{\mathcal{S}_X}$ be the set of fraglets contained by $\mathcal{S}_X$.

Definition 1. *The intruder's knowledge* $\Phi_{\mathcal{S}_X}^{\mathsf{F}_{\mathcal{S}_X}}$ *is defined as:*

$$\Phi_{\mathcal{S}_X}^{\mathsf{F}_{\mathcal{S}_X}} = \{s_i | f_i =_{\mathcal{S}_X} [t_i \ s_i \ tail_i] \in \mathsf{D}(\mathsf{F}_{\mathcal{S}_X})\}$$

for some generic auxiliary or instruction tag $t_i, i = 0, \ldots, m$, and some generic sequence of symbols $tail_i, i = 0, \ldots, m$.

Security properties: Secrecy. We give here the definition of one of the most common security properties, secrecy, within a fraglets framework.

Intuitively, a message is secret when it is only known by the parties that should share that secret. Thus, in a fraglet context, a symbol is a secret between $\mathcal{S}_S$ and $\mathcal{S}_R$ when it is not possible for $\mathcal{S}_X$ to know that symbol.

We let $\mathsf{F}_{\mathcal{S}_S}^0$ and $\mathsf{F}_{\mathcal{S}_R}^0$ to be the initial, and fixed (according to the protocol in which the honest roles are engaged), set of fraglets stored at, resp., $\mathcal{S}_S$ and $\mathcal{S}_R$, at the beginning of the computation.

Analogously, $\mathsf{F}_{\mathcal{S}_X}^0$ is the set of fraglets initially contained by $\mathcal{S}_X$. A priori, we do not make any assumption on this set, apart from the fact that it does not contain private information of the honest roles, such as private keys of $\mathcal{S}_S$ and $\mathcal{S}_R$, and their shared secret key.

Thus, the following definition dictates when the secrecy property is preserved.

Definition 2. *The secrecy property* $Sec(s)_{\mathcal{S}_X}$ *of a symbol s is preserved if* $\forall \mathsf{F}_{\mathcal{S}_X}^0$ *and*

$$\forall (\mathsf{F}_{\mathcal{S}_S}' \cup \mathsf{F}_{\mathcal{S}_X}' \cup \mathsf{F}_{\mathcal{S}_R}') \in \mathsf{D}(\mathsf{F}_{\mathcal{S}_S}^0 \cup \mathsf{F}_{\mathcal{S}_X}^0 \cup \mathsf{F}_{\mathcal{S}_R}^0) \text{ then } s \notin \Phi_{\mathcal{S}_X}^{\mathsf{F}_{\mathcal{S}_X}'}.$$

This means that, for every possible set of fraglets initially contained by the adversary's store, and for every possible union of fraglets' sets contained at $\mathcal{S}_S$, $\mathcal{S}_X$, and $\mathcal{S}_R$ that are derivable from the initial sets by applying every possible rule of the rewrite system, $\mathcal{S}_X$ will never know the secret symbol. This notion of secrecy is violated in the fraglets-based instantiation of the flawed NSPK, as shown in Section 5.

4   Specification and Execution of Fraglets in Maude

Maude is "a programming language that models (distributed) systems and the actions within those systems" [17]. The system is specified by defining algebraic data types axiomatizing system's states, and rewrite rules axiomatizing system's local transitions.

In this section we present our Maude executable specification for (crypto)fraglets. In particular, we define an algebra for them, *i.e.*, the *sorts* (types for values), and the *equationally specifiable operators* acting on those sorts (and constants). Also, we define the *rewrite laws* for describing the transitions that occur within and between the set of operators. Actually, the set of the rewrite laws represent the set of (crypto)fraglets instructions given in Tables 1, 2.

The Maude modules consisting of the core fraglets specification are basically three: FRAGLETS, FRAGLETS-RULES, and CRYPTO-FRAGLETS-RULES.

The functional module FRAGLETS provides declarations of sorts, *e.g.*, fraglets, symbols, stores, and public and private keys, and operators on those sorts, *e.g.*, concatenation of fraglets, and concatenations of fraglet stores. It also defines subsort relationships. For instance, symbols, stores, and public and private keys are understood as specialized fraglets. The module also provides reserved ground terms representing the names of the instructions (match, dup, exch, ...), and operators to encrypt fraglets, by means of either symmetric (crypt) or asymmetric (asymcrypt) encryption.[1]

```
fmod FRAGLETS is

  sort Fraglet .
  sort Key PKey . ---PKey is the sort for asymmetric cryptography
  sort Symb Store .
  sort FragletSet FragletSet@Store FragletStoreSet .

  subsort Symb Store Key PKey < Fraglet < FragletSet .
  subsort FragletSet@Store < FragletStoreSet .

  op nil : -> Fraglet .
  op _ _ : Fraglet Fraglet -> Fraglet [ctor assoc id: nil] .

  op empty : -> FragletSet .
  op _ , _ : FragletSet FragletSet -> FragletSet [assoc comm id: empty] .

  op _@_ : FragletSet Store -> FragletSet@Store .

  --- store concatenation
  op _ ; _ : FragletStoreSet FragletStoreSet -> FragletStoreSet [assoc comm ] .
  op [ _ ] : Fraglet -> FragletSet .

  op match : -> Symb .
  op dup : -> Symb .
  op exch : -> Symb   .
  ...
  ...
  op split : -> Symb .
  op send : -> Symb .

  op enc : -> Symb .
  op dec : -> Symb   .

  op fst : -> Symb .
  op snd : -> Symb .

  op crypt : Fraglet Key -> Fraglet .      --- symmetric encryption
  op asymcrypt : Fraglet PKey -> Fraglet . --- asymmetric encryption
  ...

endfm
```

---

[1] Appropriate equations for all the operators are defined in the complete specification.

The system module FRAGLETS-RULES defines the rewrite rules encoding the instructions given in Table 1. Below, we highlight (part of) the rules for dup and for a modified send instruction following the threat model defined in Section 3.

```
mod FRAGLETS-RULES is
  protecting FRAGLETS .

  vars T S : Symb .
  var TAIL : Fraglet .
  vars A B X : Store .
  vars  FS1 FS2 : FragletSet .

  rl [DUP] : [dup T S TAIL] => [T S S TAIL] .
  ...
  ...
  rl [SEND] : (([send X TAIL], FS1) @ A ) => (FS1 @ A) ; [TAIL] @ X .
  rl [SEND] : (([send X TAIL], FS1) @ A ) ;  (FS2 @ X ) => (FS1 @ A) ;
                (([TAIL], FS2) @ X) .
  ...
endm
```

CRYPTO-FRAGLETS-RULES defines the rewrite rules for cryptography and declares the pairs of public/private keys used for protocol specifications. Decryption instructions are defined as conditional rules (crl [DEC]): with symmetric keys, decryption is possible only if the key used for encryption is equal to the key that one intends to use for decryption; analogously, with asymmetric keys, decryption is possible only if the key used for encryption and the key intended to use for decryption form a pair. To this aim, auxiliary operators isKey and keypair have been equationally defined.

```
mod CRYPTO-FRAGLETS-RULES is
  protecting FRAGLETS .
  protecting FRAGLETS-RULES .

  op _ _ isKey : Key Key -> Bool .         --- aux op for dec rule
  op _ _ keypair : PKey PKey -> Bool [comm] . --- aux op for asym dec rule

  --- public and private keys declarations:
  ops pka pkb pkx : -> PKey .  --- public keys
  ops ska skb skx : -> PKey .  --- private keys

  vars    --- some variables declarations

  eq K K isKey = true .
  eq K1 K isKey = false [owise] .

  ---  which keys form a pair:
  eq pka ska keypair = true .
  eq pkb skb keypair = true .
  eq pkx skx keypair = true .
  eq K K1 keypair = false [owise] .

  --- SYMMETRIC
  rl [ENC] : [enc NewTag K TAIL] => [ NewTag crypt(TAIL,K) ] .
  rl [ENC] : [enc] => empty .
  crl [DEC] : [dec NewTag K crypt(TAIL,K1)] => [NewTag TAIL]
                if (K K1 isKey == true) .

  --- ASYMMETRIC
  rl [ENC] : [enc NewTag PK TAIL] => [ NewTag asymcrypt(TAIL,PK) ] .
  rl [ENC] : [enc NewTag SK TAIL] => [ NewTag asymcrypt(TAIL,SK) ] .
  crl [DEC] : [dec NewTag PK asymcrypt(TAIL,SK)] => [NewTag TAIL]
```

```
                      if (PK SK keypair == true) .
    crl [DEC] : [dec NewTag SK asymcrypt(TAIL,PK)] => [NewTag TAIL]
                      if (PK SK keypair == true) .
    ...
endm
```

To actually do something with those modules, one should use some strategies for applying the rules. A default strategy provided by Maude is implemented by the *rewrite* command, that explores one possible sequence of rewrites, starting by the set of rules and an initial state [16]. For example, plugging in "rew [enc newtag k tail] ." into the Maude environment, we obtain as a result "[newtag crypt(tail, k)]". The *search* command is also very convenient. *A priori*, it gives all the possible sequence of rewrites between an initial and a final state supplied by the user. Practically, since for certain systems the search could not terminate, the command is decorated with an optional bound on the number of desired solutions and on the maximum depth of the search.

5   A Case Study: Fraglets-Based NSPK

In this section, we first recall the Needham-Schroeder Public Key protocol (NSPK) [19], a paradigmatic security protocol, widely examined by protocol researchers. Considering two interleaved runs of the protocol, [12] found an attack leading to both an authentication and a secrecy failure, and supplied an amended version of the protocol.

Then, we show the Maude specification of the fraglets-based NSPK. Executing the specification, we find the secrecy attack. This starting example illustrates the usefulness of executing fraglets specifications in the Maude engine for validation purposes.

The NSPK protocol. NSPK tries to establish an authenticated communication between a pair of agents, A and B. The protocol is based on public-key cryptography: each agent possesses a pair of private/public keys. While the public key can be accessed by all agents of the distributed system, the private key should remain a secret of its owner. Also, the protocol makes use of the nonces nA, and nB. Nonces are freshly generated, random numbers, generally exploited in cryptographic protocols to assure freshness of messages. Actually, one of the intents of NSPK is also to exchange secret values nA, and nB to be used for subsequent encrypted communication between A and B.

The original version of (part of) the protocol is hereafter presented. We denote the transmission of message *msg* from sender S to receiver R as $S \rightarrow R : msg$. Also, encryption of message *msg* with public key *pk* is denoted as $\{msg\}_{pk}$.

$$1\ A \rightarrow B : \{A, nA\}_{pkB}$$
$$2\ B \rightarrow A : \{nA, nB\}_{pkA}$$
$$3\ A \rightarrow B : \ \ \{nB\}_{pkB}$$

In message 1, A sends his identity and his newly generated nonce nA to B, encrypted with the public key of B, pkB. B decrypts message 1 with the correspondent private key skB. Then, B creates her nonce nB and sends back to A the two nonces, encrypted with the public key of A, pkA (message 2). Finally, A decrypts message 2, retrieves the nonce nB, and sends it back, encrypted, to B.

Once terminated a run of this protocol, it would seem reasonable that:

1. A and B know with whom they have been interacting; indeed, A can be assured that message 2 came from B, because B is the only agent who can decrypt message 1, *i.e.*, the message sent by A and containing nA. Analogously, B can be assured of being talked to A, because A is the only agent who can decrypt message 2.
2. A and B agree on the values of nA and nB.
3. No one else knows the values of nA and nB (secrecy).

Maude specification and execution of the fraglets-based NSPK. For many years the NSPK protocol has been believed to satisfy those properties. In [12], Gavin Lowe discovered an attack, in which an adversary X acts as a honest principal with A in a first run of the protocol, while she masquerades as A for B in a second run of the protocol:

$$
\begin{array}{lll}
a.1 & A \to X : & \{A, nA\}_{pkX} \\
b.1 & X(A) \to B : & \{A, nA\}_{pkB} \\
b.2 & B \to X(A) : & \{nA, nB\}_{pkA} \\
a.2 & X \to A : & \{nA, nB\}_{pkA} \\
a.3 & A \to X : & \{nB\}_{pkX} \\
b.3 & X \to B : & \{nB\}_{pkB}
\end{array}
$$

where $X(A)$ represents X generating (resp. receiving) the message, making it appear as generated (resp. received) by A. What happens is that A starts Session a with X, that, in its turn, starts Session b with B, pretending to be A. At the end of the two sessions, B thinks that i) she has been communicating with A, while this is not the case, and ii) she and A share exclusively nA and nB, while this is not the case.

Module NSPK-flawed-intruder declares the names of the stores, the nonces and the auxiliary tags necessary to make fraglets opportunely react with each other according to NSPK.

```
mod NSPK-flawed-intruder is
    protecting FRAGLETS .
    protecting FRAGLETS-RULES .
    protecting CRYPTO-FRAGLETS-RULES .

    --- two interleaved runs
    --- we consider three participants, A, B, and X.
    --- secrecy attack on Nb: at the end of the runs, B is convinced that
        Nb is a secret known only by B and A, while X knows Nb.

    ops Sa Sb Sx : -> Store .
    ops na nb : -> Key .

    --- all the auxiliary tags for fraglets transformations and reactions
    ops key key1 key2 key3   : -> Symb .
    ops auxtag auxtag1 auxtag2 auxtag3 auxtag4 auxtag5 auxtag6 : -> Symb .
    ops msga1 msga2 msga3 msga12 msga23 msgb1 msgb2 msgb3 msgb12 msgb23
        kmsga1 kmsga2 kmsga3  kmsgb1 kmsgb2 kmsgb3 secretb : -> Symb .
endm
```

We can explore one possible sequence of rewrites by running the *rewrite* command. The argument of *rewrite* is the initial configuration of the three stores $\mathcal{S}_A$, $\mathcal{S}_B$, and $\mathcal{S}_X$. Actually, this configuration represents the fraglets program to execute two interleaved sessions of NSPK. Figure 2 shows the screenshot of executing the fraglets-based NSPK.

```
File  Edit  View  Terminal  Help
Maude> load NSPK-flawed-intruder-secrecy-nb.maude
==============================================
rewrite in NSPK-flawed-intruder : (([key pkx],[key2 ska],[msga1 na Sa],[matchs
    key match msga1 enc auxtag1],[match auxtag1 send Sx kmsga1],[matchs key2
    match kmsga2 dec auxtag3],[match auxtag3 fst msga23],[matchs msga23 msga3],
    [match auxtag4 send Sx kmsga3],[matchs key match msga3 enc auxtag4]) @ Sa)
    ; (([key skb],[key1 pka],[matchs key match kmsgb1 dec auxtag3],[match
    auxtag3 fst msgb12],[matchs msgb12 msgb2 nb],[match auxtag4 send Sx
    kmsgb2],[matchs key1 match msgb2 enc auxtag4]) @ Sb) ; ([key skx],[key1
    pka],[key2 pkb],[matchs key match kmsga1 dec msgb1],[matchs key2 match
    msgb1 enc auxtag2],[match auxtag2 send Sb kmsgb1],[match kmsgb2 send Sa
    kmsga2],[matchs key match kmsga3 dec auxtag5],[match auxtag6 send Sb
    kmsgb3],[matchs key2 matchs auxtag5 enc auxtag6]) @ Sx .
rewrites: 53 in 4ms cpu (3ms real) (13250 rewrites/second)
result FragletStoreSet: (([key pkx],[key2 ska],[msga23 nb]) @ Sa) ; (([key
    skb],[key1 pka],[msgb12 na],[kmsgb3 asymcrypt(nb, pkb)]) @ Sb) ; ([key
    skx],[key1 pka],[key2 pkb],[auxtag5 nb]) @ Sx
==============================================
```

Fig. 2. Fraglets-based NSPK: execution in Maude

At the end of the computation, the result is of sort FragletStoreSet (*i.e.*, a set of fraglet stores). In particular, S_A contains the newly received nonce nb. S_B contains nonce na, received in message b1, and the last encrypted message received by S_A (with her nonce nb). Finally, the adversary's store contains fraglet [$auxtag5$ nb], leading to a secrecy attack with respect to nb.

We can also use *search*, looking for more than one possible sequence of rewrites from the initial configuration in the screenshot to a state where [$auxtag5$ nb] belongs to S_X. The result gives one of such sequences obtained after 13422 rewrites in 880 milliseconds (cpu).

6   Conclusions

In this paper, we defined a threat model for a refined version of cryptofraglets. On that model, it is possible to define security properties for fraglets, starting from the notion of the adversary's knowledge. As an example, we dealt with the secrecy property. Then, we proposed an executable specification of cryptofraglets in Maude, together with an executable specification of a fraglets-based version of the well known security protocol NSPK. A security verification was achieved with respect to secrecy.

On the one hand, this illustrates the usefulness of executing fraglets specifications in the Maude engine for validation purposes. However, this represents only a partial verification on a reduced scenario. Indeed, we focused on a fixed initial set of fraglets in the adversary store. Furthermore, the sets of fraglets derivable from the initial sets by applying every possible rule of the rewrite system is fixed.

One of the main goals for the future is to extend this work to more comprehensive scenarios, in order to verify the secrecy property as defined in section 3, as well as other kinds of security properties, in a more exhaustive way. Indeed, it is possible to express strategies for executing Maude specifications in Maude itself (see, *e.g.*, [6] for strategies examples). As future work, we intend to use some ad-hoc defined strategies for extending our verification framework.

Acknowledgments. This work is partly supported by the FP6-027748 EU project BIONETS (*BIOlogically-inspired autonomic NETworks and Services*).

The authors would like to thank Roberto Bruni and Alberto Lluch Lafuente for their kind support on Maude matters, Grit Denker for providing pointers to useful Maude specifications, and Daniel Schreckling for valued discussions on the cryptofraglets syntax.

References

1. Van Baalen, J., Böhne, T.: Automated protocol analysis in Maude. In: Hinchey, M.G., Rash, J.L., Truszkowski, W.F., Rouff, C.A., Gordon-Spears, D.F. (eds.) FAABS 2002. LNCS (LNAI), vol. 2699, pp. 68–78. Springer, Heidelberg (2003)
2. The BIONETS website, http://www.bionets.eu/
3. Cervesato, I., Durgin, N., Lincoln, P.D., Mitchell, J.C., Scedrov, A.: A meta-notation for protocol analysis. In: Proc. CSFW-12, pp. 55–69. IEEE, Los Alamitos (1999)
4. Clarke, E., Jha, S., Marrero, W.: Verifying security protocols with Brutus. ACM Transactions on Software Engineering and Methodology 9(4), 443–487 (2000)
5. Clavel, M., Durán, F., Eker, S., Lincoln, P., Martí-Oliet, N., Meseguer, J., Talcott, C.: All About Maude - A High-Performance Logical Framework. LNCS, vol. 4350. Springer, Heidelberg (2007)
6. Denker, G., Meseguer, J., Talcott, C.: Protocol specification and analysis in Maude. In: Formal Methods and Security Protocols (1998)
7. Dolev, D., Yao, A.: On the security of public key protocols. IEEE Trans. Inf. Theory 29(2), 198–208 (1983)
8. Egidi, L., Petrocchi, M.: Modelling a secure agent with team automata. In: Proc VODCA 2004. ENTCS, pp. 119–134. Elsevier, Amsterdam (2005)
9. Focardi, R., Martinelli, F.: A uniform approach for the definition of security properties. In: Wing, J.M., Woodcock, J.C.P., Davies, J. (eds.) FM 1999. LNCS, vol. 1708, pp. 794–813. Springer, Heidelberg (1999)
10. FRAGLETS website (last access: 18/06/2009), http://www.fraglets.net
11. Lenzini, G., Gnesi, S., Latella, D.: Spider: a Security Model Checker. In: Proc. FAST 2003, pp. 163–180 (2003); Also, Technical Report ITT-CNR-10, Informal proceedings (2003)
12. Lowe, G.: Breaking and fixing the needham-schroeder public-key protocol using fdr. In: Margaria, T., Steffen, B. (eds.) TACAS 1996. LNCS, vol. 1055, pp. 147–166. Springer, Heidelberg (1996)
13. Lynch, N.: I/O automaton models and proofs for shared-key communication systems. In: Proc. CSFW 1999, pp. 14–31. IEEE, Los Alamitos (1999)
14. Martinelli, F., Petrocchi, M.: Access control mechanisms for fraglets. In: BIONETICS, ICST (2007)
15. Martinelli, F., Petrocchi, M.: Signed and weighted trust credentials for fraglets. In: BIONETICS, ICST (2008)
16. Maude System website (last access: 18/06/2009), http://maude.cs.uiuc.edu/
17. McCombs, T.: Maude 2.0 Primer (2003), http://maude.cs.uiuc.edu/download/
18. Meseguer, J.: Research directions in rewriting logic. In: Computational Logic. LNCS, vol. 165, Springer, Heidelberg (1997)
19. Needham, R., Schroeder, M.: Using encryption for authentication in large networks of computers. Communications of the ACM (21), 393–399 (1978)

20. Petrocchi, M.: Formal techniques for modeling and verifying secure procedures. PhD thesis, University of Pisa (May 2005)
21. Petrocchi, M.: Crypto-fraglets. In: BIONETICS. IEEE, Los Alamitos (2006)
22. Tschudin, C.: Fraglets - a metabolistic execution model for communication protocols. In: Proc. AINS 2003 (2003)
23. Tschudin, C., Yamamoto, L.: A metabolic approach to protocol resilience. In: Smirnov, M. (ed.) WAC 2004. LNCS, vol. 3457, pp. 191–206. Springer, Heidelberg (2005)
24. Yamamoto, L., Tschudin, C.: Experiments on the automatic evolution of protocols using genetic programming. In: Stavrakakis, I., Smirnov, M. (eds.) WAC 2005. LNCS, vol. 3854, pp. 13–28. Springer, Heidelberg (2006)
25. Yamamoto, L., Tschudin, C.: Genetic evolution of protocol implementations and configurations. In: Proc. SelfMan 2005 (2005)

Ensuring Fast Adaptation in an Ant-Based Path Management System

Laurent Paquereau and Bjarne E. Helvik

Centre for Quantifiable Quality of Service in Communication Systems⋆,
Norwegian University of Science and Technology, Trondheim, Norway
{laurent.paquereau,bjarne}@q2s.ntnu.no

Abstract. The Cross-Entropy Ant System (CEAS) is an Ant Colony
Optimization (ACO) system for distributed and online path management
in telecommunication networks. Previous works on CEAS have focused
on reducing the overhead induced by the continuous sampling of paths. In
particular, elite selection has been introduced to discard ants that have
sampled poor quality paths. This paper focuses on the ability of the
system to adapt to changes in dynamic networks. It is shown that not
returning ants may cause stagnation as that tends to make stale states
persist in the network. To mitigate this undesirable side-effect, a novel
pheromone trail evaporation strategy, denoted Selective Evaporation on
Forward (SEoF), is presented. By allowing ants to decrease pheromone
trail values on their way forward, it enforces a local re-opening of the
search process in space upon change when elite selection is applied.

Keywords: CEAS, elite selection, Selective Evaporation on Forward.

1 Introduction

Ant Colony Optimization (ACO) [1] systems are systems inspired by the foraging
behaviour of ants and designed to solve discrete combinatorial optimization problems. More generally, ACO systems belong to the class of Swarm Intelligence (SI)
systems [2]. SI systems are formed by a population of agents, which behaviour is
governed by a small set of simple rules and which, by their collective behaviour,
are able to find good solutions to complex problems. ACO systems are characterized by the indirect communication between agents - (artificial) ants - referred
to as stigmergy and mediated by (artificial) pheromones. In nature, pheromones
are a volatile chemical substance laid by ants while walking that modifies the
environment perceived by other ants. ACO systems have been applied to a wide
range of problems [1]. The Cross-Entropy Ant System (CEAS) is such a system
for path management in dynamic telecommunication networks.

The complexity of the problem arises from the non-stationary stochastic dynamics of telecommunication networks. A path management system should adapt

---

⋆ "Centre for Quantifiable Quality of Service in Communication Systems, Centre of
Excellence" appointed by The Research Council of Norway, funded by the Research
Council, NTNU, UNINETT and Telenor. http://www.q2s.ntnu.no

E. Altman et al. (Eds.): Bionetics 2009, LNICST 39, pp. 24–35, 2010.

to changes including topological changes, e.g. link/node failures and restorations, quality changes, e.g. link capacity changes, and traffic pattern changes. The type, degree and time-granularity of changes depend on the type of network. For instance, the level of variability in link quality is expected to be higher in a wireless access network than in a wired core network.

Generally, the performance of an ACO system is related to the number of iterations required to achieve a given result. Specific to the path management problem in telecommunication networks are the additional requirements put on the system in terms of time and overhead. On changes, the system should adapt, i.e. converge to a new configuration of paths, in short time and with a small overhead. In addition, finding a good enough solution in short time is at least as important as finding the optimal solution, and there is a trade-off between quality of the solution, time and overhead.

Previous works on CEAS have focused on reducing the overhead induced by the continuous sampling of solutions [3]. In particular, *elite selection* has been introduced in [4] to reduce the overhead by allowing only "good" ants to update pheromone trails. The present work addresses the adaptivity of the system. ACO systems are intrinsically adaptive and this characteristic has been used as an argument for applying ACO algorithms to dynamic problems. ACO systems have been shown to be able to adapt to changes and techniques have been developed to prevent from *stagnation*[1] [5]. However, to our knowledge, the adaptivity of ACO systems in itself, for instance in terms of number of iterations needed to converge after a change, has received little attention.

In this paper, a novel extension, denoted *Selective Evaporation on Forward* (SEoF), is introduced to improve the adaptivity of the system. The rest of this paper is organized as follows. Section 2 provides a brief introduction to CEAS. For a comprehensive presentation, the reader is referred to [3]. Section 3 describes elite selection and Section 4 characterizes the stagnation caused by not returning ants. Next, Section 5 presents the SEoF extension. Finally, Section 6 discusses related work and Section 7 concludes.

2   CEAS in a Nutshell

CEAS is an asynchronous and distributed ACO system for path management in telecommunication networks based on the Cross-Entropy (CE) method for stochastic optimization [6]. Ants cooperate to collectively find and maintain minimal cost paths, or sets of paths, between source and destination pairs. Each ant performs a random search directed by the pheromone trails to find a path to a destination. Each ant also deposits pheromones so that the pheromone trails reflect the knowledge acquired by the colony thus enforcing the stigmergic behaviour characterizing ACO systems. Similarly to what happens in nature, good solutions emerge as the result of the iterative indirect interactions between ants.

---

[1] *Stagnation* refers to a system that fails to adapt, or adapts very slowly, because it has converged too hard to a solution.

Formally, let a network be represented by a bidirectional weighted graph $\mathbf{G} = (\mathbf{V}, \mathbf{E})$ where $\mathbf{V}$ is the set of vertices (nodes) and $\mathbf{E}$ the set of edges (links). $(v, i) \in \mathbf{E}$ denotes the link connecting node v to node i and $L((v, i))$ is the weight (cost) of link (v, i). Starting from a node s, an ant incrementally builds a path to a destination node d by moving through a sequence of neighbour nodes applying at each node a stochastic decision policy depending on the local pheromone trails and the ant internal state (*biased exploration*). At node v, the probability that an ant decides to move to node i is given by the *random proportional rule*

$$p_{t_v, vi}^{(s,d)} = \frac{\tau_{t_v, vi}^{(s,d)}}{\sum_{j \in \mathcal{N}_v} \tau_{t_v, vj}^{(s,d)}}, \ \forall i \in \mathcal{N}_v \tag{1}$$

where $\tau_{t_v, vi}^{(s,d)}$ is the pheromone trail value at node v for the link (v, i) after t_v pheromone deposits at node v, see below, and $\mathcal{N}_v \subseteq \mathbf{N}_v = \{i \in \mathbf{V} \mid (v, i) \in \mathbf{E}\}$ is the set of neighbours of node v not yet visited by the ant. After it has reached its destination d, the ant backtracks. On its way backward, it triggers pheromone evaporation[2]

$$\tau_{t_v-1, vi}^{(s,d)} \leftarrow \beta \cdot \tau_{t_v-1, vi}^{(s,d)}, \ \forall i \in \mathbf{N}_v, \ \forall v \in \boldsymbol{\pi}_{t, [s,d]} \tag{2}$$

and deposits pheromones (*online delayed* pheromone release)

$$\tau_{t_v, vi}^{(s,d)} \leftarrow \tau_{t_v-1, vi}^{(s,d)} + I\left((v, i) \in \boldsymbol{\omega}_{t, [s,d]}\right) \cdot \Delta\tau_t^{(s,d)}, \ \forall i \in \mathbf{N}_v, \ \forall v \in \boldsymbol{\pi}_{t, [s,d]} \tag{3}$$

where $\beta \in [0, 1)$ denotes the *memory factor*, $\boldsymbol{\pi}_{t, [s,d]} = \langle s, v_1, v_2, \ldots, v_{h-1}, d \rangle$ the sequence of nodes traversed by the ant, $\boldsymbol{\omega}_{t, [s,d]} = \langle (s, v_1), (v_1, v_2), \ldots, (v_{h-1}, d) \rangle$ the sequence of links, $\Delta\tau_t^{(s,d)}$ the amount of pheromones deposited (*pheromone increment*), $I(x) = 1$ if x is true, 0 otherwise, and t is the number of ants that have returned from d[3]. $\Delta\tau_t^{(s,d)}$ is chosen so that (1) minimizes the cross-entropy between two consecutive sets of random proportional rules $\mathbf{p}_t^{(s,d)} = \{p_{t, vi}^{(s,d)}\}_{\forall v, i}$ and $\mathbf{p}_{t-1}^{(s,d)}$ subject to the cost history $\mathbf{L}_{t, [s,d]} = \{L(\boldsymbol{\omega}_{k, [s,d]}) \mid k = 1, \ldots, t\}$, where $L(\boldsymbol{\omega}_{k, [s,d]}) = \sum_{\forall (i,j) \in \boldsymbol{\omega}_{k, [s,d]}} L((i, j))$ is the cost of the path $\boldsymbol{\omega}_{k, [s,d]}$.

$$\Delta\tau_t^{(s,d)} = H\left(L(\boldsymbol{\omega}_t), \gamma_t^{(s)}\right) = e^{-L(\boldsymbol{\omega}_t)/\gamma_t^{(s)}} \tag{4}$$

where $\gamma_t^{(s)}$ is an internal parameter called the *temperature* and determined at d by minimizing it subject to $h_t(\gamma_t) \geqslant \rho$. $h_t(\gamma_t) = \beta \cdot h_{t-1}(\gamma_t) + (1-\beta) \cdot H(L(\boldsymbol{\omega}_t), \gamma_t)$ is the overall auto-regressive performance function and $\rho \in (0, 1)$ a configuration parameter (*search focus*). In the following, the subscripts and superscripts referencing the source and destination nodes are omitted to help readability.

---

[2] More precisely, $\tau_{t, vi}$ is implemented as an auto-regressive function of γ_t and evaporation is applied on the auto-regressive variables. See for instance [3].

[3] t is incremented when an ant is at its destination, hence the number of pheromone deposits at node v is $t_v = \sum_{k=1}^{t} I((v, \cdot) \in \boldsymbol{\omega}_{k, [s,d]})$.

The temperature γ_t controls the weights given to solutions. For a given temperature, the lower the cost, the larger the pheromone increment. For a given cost value, the lower the temperature, the smaller the pheromone increment, but the larger the relative difference with the increment for a solution of higher cost, see Figure 1. γ_t is self-adjusting. When the network conditions stay unchanged, γ_t asymptotically converges to $\tilde{\gamma}_t$. If network conditions between s and d degrade, e.g. if the quality of the best path is altered so that it is no longer the best or if the best path is no longer available, $\tilde{\gamma}_t$ becomes larger and γ_t gradually increases (*reheating change*). If the network conditions between s and d improve, e.g. when a better path has become available and has been discovered, $\tilde{\gamma}_t$ becomes smaller and γ_t gradually decreases (*cooling change*).

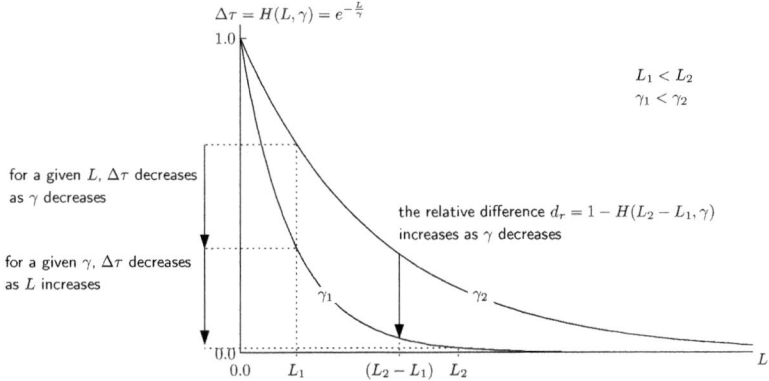

Fig. 1. Changes in pheromone increments as cost and temperature vary

To bootstrap the system, ants do not apply (1) but a *uniformly distributed proportional rule* (*uniform exploration*)

$$p_{t_v,vi} = \frac{1}{|\mathcal{N}_v|}, \ \forall i \in \mathcal{N}_v \tag{5}$$

During normal operation, a given percentage of ants applying (5) is maintained. The purpose of such *explorer ants* is threefold: (i) to ensure that new solutions are discovered, (ii) to prevent the system from converging too hard, and (iii) to maintain sparse pheromone trails on alternative solutions providing roughly up-to-date bootstrapping information in case of a change.

3   Elite Selection

Elite selection consists in only letting ants that have sampled relatively good paths (*elite paths*), i.e. paths which cost is below a certain cut-off level (*elite selection level*), update the temperature and backtrack. The rationale is that relatively poor quality solutions, i.e. which cost is beyond this level, lead to negligible changes in the pheromone trail distributions and, hence, that ants that have sampled those paths can be discarded at the destination without affecting the performance of the system otherwise. Elite selection is therefore primarily an overhead reduction technique. However, it also contributes to improving the convergence speed of the system in terms of number of iterations by focusing the search around the best solutions [4].

Formally, let n be the total number of ants arrived at d from s (*forward ants*), ω_k^* the path followed by the kth ant, and $\boldsymbol{\Omega}_n^*$ the set of all candidate paths. Elite selection can be formulated as

$$\omega_n^* \in \hat{\boldsymbol{\Omega}}_n \Leftrightarrow L(\omega_n^*) \leqslant \chi_n \tag{6}$$

where $\hat{\boldsymbol{\Omega}}_n \subseteq \boldsymbol{\Omega}_n^*$ is the set of elite paths and χ_n the elite selection level. It is shown in [4] that an appropriate elite selection level is

$$\chi_n = -\gamma_n^* \cdot \ln \rho \tag{7}$$

where γ_n^* is the temperature calculated from the total cost history $\mathbf{L}_t^* = \{L(\omega_k^*) \mid k = 1, \ldots, n\}$. Contrary to the *total temperature* γ_n^*, the *elite temperature* $\gamma_{\hat{t}}$ used in (4) to determine $\Delta\tau_t$ is computed from $\hat{\mathbf{L}}_{\hat{t}} = \{L(\hat{\omega}_k) \mid k = 1, \ldots, \hat{t}\}$ where $\hat{t} \leqslant n$ is the number of elite ants, i.e. ants that have sampled an elite path. χ_n is self-adjusting; γ_n^* adjusts to the network conditions and so does χ_n.

Ants that do not meet the elite selection criterion (6) are discarded, with the exception of explorer ants that are always returned. Hence, the number of ants that backtrack (*backward ants*) is $t \in [\hat{t}, n]$. All the operations performed at the destination are summarized in the flow chart shown in Figure 2.

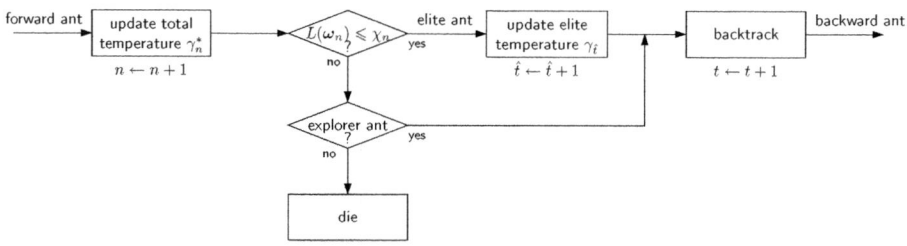

Fig. 2. Elite selection flow chart

4   Stagnation Caused by Non Returning Ants

An ant may not return because of elite selection or accidentally (loss). Looking at the adaptivity of the system, this latter case is similar to the case of an ant discarded by elite selection after the degradation of a path. Hence, it is not further considered in the following.

As long as the highest pheromone values correspond to the best path, the search process is correctly biased and elite selection is an advantageous feature as only good paths get reinforced. When the highest pheromone values do not correspond to the best path anymore, i.e. when the search process has become incorrectly biased, e.g. after a degradation of the path, it turns out to be harmful because it tends to keep stale pheromone trails longer. The crux of the problem is that pheromone trails are only updated by backward ants. Discarding ants at the destination results in no update of the pheromone values.

In particular, stagnation occurs on reheating changes because: (i) the elite set may be temporarily empty after the change, in which case even an ant following the new optimal solution does not meet the elite selection criterion $(\hat{\Omega}_n = \emptyset \Leftrightarrow L(\omega_n^*) > \chi_n, \forall \omega_n^* \in \Omega_n^*)$, and (ii) the elite temperature is lower than what it will be when the system has converged so pheromone deposits are smaller than what they will be when the system has converged $(\gamma_t < \tilde{\gamma}_t \Leftrightarrow \Delta\tau_t < \tilde{\Delta}\tau_t = H(L(\omega_t), \tilde{\gamma}_t))$. On cooling changes, even though the search process becomes incorrectly biased, elite selection does not cause stagnation because: (i) the elite set is never empty so pheromone trails can always potentially be updated $(\hat{\Omega}_n \neq \emptyset)$, (ii) the elite temperature is higher than what it will be when the system has converged so pheromone deposits are larger than what they will be when the system has converged $(\gamma_t > \tilde{\gamma}_t \Leftrightarrow \Delta\tau_t > \tilde{\Delta}\tau_t)$, and (iii) elite selection focuses the search on the best paths and thus prevents temperature increase on sampling low quality solutions.

The above cases are demonstrated by simulation of a simple scenario chosen for illustration. CEAS, with and without elite selection, is applied to find the shortest path between nodes s and d in the four node network shown to the left in Figure 3. At $t = 0$, all the links are operational. The link (a, b) is taken down at $t = 3000$ [s] and restored at $t = 11000$ [s]. Node s generates ants at rate $\lambda = 1$ [ant/s] out of which 5% are explorer ants. The path memory factor applied is $\beta = 0.998^4$, and the search focus $\rho = 0.01$. Figure 3 shows the probability $p_{t,sa}$ of forwarding ants to node a at node s. The results are averaged over 30 replications. For $t < 3000$ [s] and $t \geqslant 11000$ [s], the best path is $\langle s, a, b, d \rangle$. Hence,

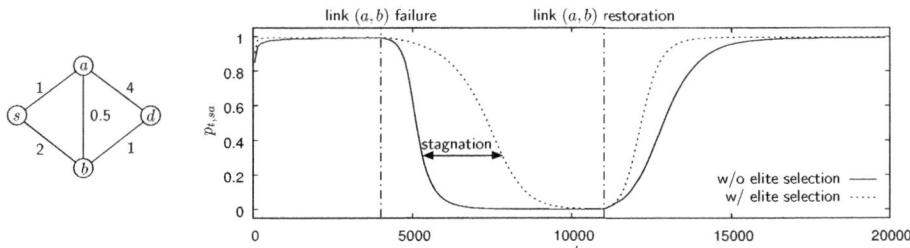

Fig. 3. Characterization of the stagnation caused by elite selection

---

[4] Such a high β is chosen to emphasize the stagnation phenomenon but is unnecessary to solve such a simple problem.

when the system has converged, $p_{t,sa} \approx 1$. In both intervals, the temperature decreases and using elite selection leads to faster convergence. When (a, b) is down, the best path is $\langle s, b, d \rangle$, hence, when the system has converged, $p_{t,sa} \approx 0$. Immediately after the link break, the search process is incorrectly biased and the temperature should increase, and elite selection causes stagnation[5].

5   Selective Evaporation on Forward

Stagnation occurs when the search process becomes wrongly biased because most of the ants are discarded at the destination and no update is triggered at intermediate nodes. However, the fact that an ant does not backtrack is an information in itself; it means that the sampled path is not, or no longer, an elite path. When the system has converged, the probability that an ant follows an elite path is high. Hence, if the network conditions are stable, the probability that an ant returns is also high. Therefore, if the system has converged, for a node along the sampled path, not receiving a backward ant indicates with a high probability that network conditions have changed. The idea is to exploit this knowledge locally at each node to reflect the change on the pheromone distribution and mitigate stagnation.

The approach followed, denoted *Selective Evaporation on Forward* (SEoF), is a hybrid online pheromone evaporation strategy combining step-by-step evaporation on the selected links and delayed evaporation on the other links. Formally, a forward ant triggers evaporation at node v on the selected link (v, i)

$$\tau_{t_v, vi} \leftarrow \beta \cdot \tau_{t_v, vi}, \tag{8}$$

and, on its way backward, (2) is replaced by

$$\tau_{t_v - 1, vi} \leftarrow \beta \cdot \tau_{t_v - 1, vi}, \ \forall i \in \mathbf{N}_v \mid (v, i) \notin \boldsymbol{\omega}_t, \ \forall v \in \boldsymbol{\pi}_t. \tag{9}$$

If an ant samples an elite path, the behaviour of the system is unmodified. If an ant samples a path that is not in the elite set, at each node along the path the pheromone trail value on the selected link only is reduced, so the probability that an ant samples the same path again is decreased. When the system has converged, the effect of this extension is marginal since most of the ants follow an elite path. On the other hand, when the pheromone distribution becomes wrongly biased after a change, the probability that an ant follows a path that is not in the elite set is high and gradually decreasing the pheromone trail values along this path does make a difference. Figure 4 shows how this strategy, denoted 'SEoF plain', effectively mitigates stagnation on reheating changes when applied to the scenario used in Section 3. In addition, in this case, the optimal solution is easy to find and reducing the pheromone trail values on paths that are not in the elite set also improves the performance of the system in cooling phases by accentuating the focus on the best solution.

Now, when the system has not yet converged, there is a non-negligible probability that an ant follows a path that is not in the elite set and therefore locally

---

[5] In this context, stagnation means that the number of iterations needed to converge is greater than the number of iterations that is needed if elite selection is not used.

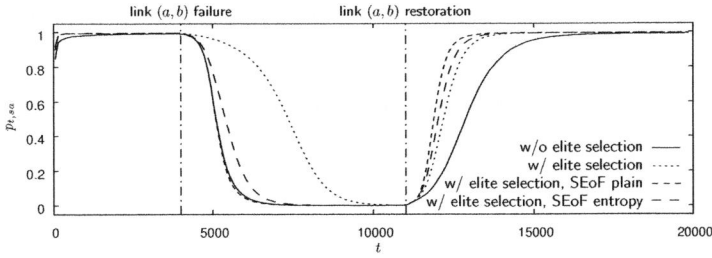

Fig. 4. Mitigating stagnation caused by elite selection using SEoF

reduces the probability of choosing a link although it may be part of a good path. Hence, this simple strategy may divert the system from good solutions and result in a slower convergence and/or convergence to a poorer solution, especially when good solutions are hard to find. See Appendix A for an illustration. To avoid this potential pitfall, (8) is replaced by

$$\tau_{t_v,vi} \leftarrow (\beta + (1-\beta)E_{t_v,v}) \cdot \tau_{t_v,vi} \tag{10}$$

and (9) by

$$\tau_{t_v-1,vi} \leftarrow \begin{cases} \beta \cdot \tau_{t_v-1,vi}, \ \forall i \in \mathbf{N}_v \mid (v,i) \notin \boldsymbol{\omega}_t, \ \forall v \in \boldsymbol{\pi}_t \\[2mm] \dfrac{\beta}{\beta + (1-\beta)E_{t_v-1,v}} \cdot \tau_{t_v-1,vi}, \ (v,i) \in \boldsymbol{\omega}_t, \ \forall v \in \boldsymbol{\pi}_t \end{cases} \tag{11}$$

where

$$E_{t_v,v} = \frac{-\sum_{\forall i \in \mathbf{N}_v} p_{t_v,vi} \log p_{t_v,vi}}{\log |\mathbf{N}_v|} \tag{12}$$

denotes the (normalized) entropy at node v.

The entropy reflects how open is the search process at node v. The idea behind this revised scheme is to balance between forward and backward evaporation on selected links depending on how severe the stagnation would be if the search process was wrongly biased. Stagnation may only occur if $E_{t_v,v} < 1$ and is all the more so severe as $E_{t_v,v}$ is low. If the entropy is close to 0 and the pheromone distribution is wrongly biased, it enforces a strong local re-opening of the search process. If the system has not yet converged, the entropy at node v will still be relatively high and evaporation will occur mostly on the way back thus alleviating the diversion effect caused by SEoF. Compared to 'SEoF plain', this revised scheme, denoted 'SEoF entropy', leads to a slightly slower convergence in the case of the scenario used in Section 3 because the pheromone reduction is lessened as the search process is re-opened. Nevertheless, it still significantly mitigates the stagnation caused by elite selection. See Figure 4. Applied to a more complex problem, it effectively reduces the diversion effect caused by SEoF, see Appendix A. More generally, SEoF is shown to improve the performance of CEAS on reheating changes as illustrated in the Appendix B.

SEoF has the following attractive characteristics: it is an online, distributed, self-adjusting, gradual and problem-independent approach, it is based on local information only, and it does not require any extra configuration parameter. Moreover, it retains and makes use of the available information about alternative solutions.

6   Related Work

Pheromone trail evaporation and elitism are not specific to CEAS. Evaporation is a core component of the ACO meta-heuristic allowing a colony to forget about old solutions and integrated in most ACO systems[6]. Elitism has been proposed for static optimization problems to improve the convergence speed by reinforcing the best solution [9,10,11]. However, to the best of our knowledge, selective evaporation is a novel idea and self-adjusting elite selection as it is implemented in CEAS remains original.

Ant Colony System (ACS) [10] also uses a hybrid pheromone update strategy including an online step-by-step reduction of the pheromone trail values on the selected edges. However, the purpose of that reduction is radically different from what is presented in this paper. It has been introduced to counterbalance a strong elitist selection. Historically though, it is interesting to note that delayed pheromone update was early preferred to step-by-step pheromone update (AS [12]). Later, elitism was introduced to improve the performance of the system (Elitist AS [9]), before a hybrid solution combining step-by-step and delayed pheromone update was proposed to cope with drawbacks introduced by elitism.

Strategies for modifying pheromone trail values after a change are found in studies on applying AS to the dynamic Traveling Salesman Problem (TSP). For instance, *pheromone shaking* is a centralized mechanism introduced in [13] to smoothen pheromone distributions when the cost between nodes has changed and equalization strategies to adapt to node insertion/deletion are proposed in [14]. In this context, changes are globally known. Pheromone modifications are applied only once, offline (daemon activity, i.e. not triggered by ants), immediately after a change and affect all pheromone trail values at all nodes.

Other ACO systems for path management in dynamic networks also integrate mechanisms to adapt to changes, but all use problem-specific measures. AntNet [7], for instance, relies on local problem-specific heuristics, AntHocNet [8] on mechanisms borrowed from traditional MANET routing protocols, e.g. explicit route error notifications.

Finally, SEoF enforces a re-opening of the search process in space when elite selection is used. As mentioned in the introduction, another dimension is time. Self-Tuned Ant Rate (STAR) [15] is a complementary extension exploiting elite selection to improve the adaptivity of the system in terms of time to converge after a change by allowing a temporary increase of the rate of forward ants.

---

[6] Notable exceptions are ACO systems proposed for routing in telecommunication networks including AntNet [7] and AntHocNet [8]. Although the term is used for AntNet in [1], authors refers to normalization, not evaporation.

7   Conclusion

This paper introduces a selective pheromone trail evaporation strategy improving the ability of CEAS to adapt to changes. Combined with elite selection, it results in an explicit pheromone trail reduction on changes which enforces a local re-opening of the search process in space and effectively mitigates the stagnation caused by not returning ants otherwise. To our knowledge, such an approach addressing the adaptivity of an ACO system is original and it is foreseen that the principles are applicable to other ACO systems.

References

1. Dorigo, M., Di Caro, G., Gambardella, L.M.: Ant Algorithms for Discrete Optimization. Artificial Life 5(2), 137–172 (1999)
2. Bonabeau, E., Dorigo, M., Theraulaz, G.: Swarm Intelligence: From Natural to Artificial Systems. Oxford University Press, Oxford (1999)
3. Heegaard, P.E., Wittner, O.J.: Overhead Reduction in Distributed Path Management System. In: Computer Networks. Elsevier, Amsterdam (in press, 2009)
4. Heegaard, P.E., Wittner, O.J., Nicola, V.F., Helvik, B.E.: Distributed Asynchronous Algorithm for Cross-Entropy-Based Combinatorial Optimization. In: Rare Event Simulation and Combinatorial Optimization (RESIM/COP 2004), Budapest, Hungary (2004)
5. Sim, K.M., Sun, W.H.: Ant Colony Optimization for Routing and Load-Balancing: Survey and New Directions. IEEE Transactions on Systems, Man and Cybernetics, Part A: Systems and Humans 33(5), 560–572 (2003)
6. Rubinstein, R.Y.: The Cross-Entropy Method for Combinatorial and Continuous Optimization. Methodology and Computing in Applied Probability 1(2), 127–190 (1999)
7. Di Caro, G., Dorigo, M.: AntNet: Distributed Stigmergetic Control for Communications Networks. Journal of Artificial Intelligence Research 9, 317–365 (1998)
8. Ducatelle, F.: Adaptive Routing in Ad Hoc Wireless Multi-hop Networks. PhD thesis, University of Lugano, Switzerland (2007)
9. Dorigo, M., Maniezzo, V., Colorni, A.: Ant System: Optimization by a Colony of Cooperating Agents. IEEE Transactions on Systems, Man, and Cybernetics - Part B 26(1), 29–41 (1996)
10. Dorigo, M., Gambardella, L.M.: Ant Colony System: A Cooperative Learning Approach to the Traveling Salesman Problem. IEEE Transactions on Evolutionary Computation 1(1), 53–66 (1997)
11. Stützle, T., Hoos, H.H.: $\mathcal{MAX} - \mathcal{MIN}$ Ant System. Journal of Future Generation Computer Systems 16, 889–914 (2000)
12. Colorni, A., Dorigo, M., Maniezzo, V.: Distributed Optimization by Ant Colonies. In: 1st European Conference on Artificial Life (ECAL 1991), pp. 134–142 (1991)
13. Eyckelhof, C.J., Snoek, M.: Ant Systems for a Dynamic TSP. In: Dorigo, M., Di Caro, G.A., Sampels, M. (eds.) ANTS 2002. LNCS, vol. 2463, pp. 88–99. Springer, Heidelberg (2002)
14. Guntsch, M., Middendorf, M.: Pheromone Modification Strategies for Ant Algorithms Applied to Dynamic TSP. In: Boers, E.J.W., Gottlieb, J., Lanzi, P.L., Smith, R.E., Cagnoni, S., Hart, E., Raidl, G.R., Tijink, H. (eds.) EvoWorkshops 2001. LNCS, vol. 2037, pp. 213–222. Springer, Heidelberg (2001)

15. Heegaard, P.E., Wittner, O.J.: Self-tuned Refresh Rate in a Swarm Intelligence Path Management System. In: de Meer, H., Sterbenz, J.P.G. (eds.) IWSOS 2006. LNCS, vol. 4124, pp. 148–162. Springer, Heidelberg (2006)

Appendix A: Diversion Effect Caused by SEoF

This appendix illustrates the possible diversion effect caused by SEoF and its mitigation using the entropy of the local pheromone distributions. This effect is more pronounced when good solutions are hard to find. Hence, it is demonstrated by applying CEAS to solve the `fri26` symmetric static TSP taken from TSPLIB[7] and also used in [4]. Note that CEAS has not been specifically designed to solve the TSP. Such a hard (NP-complete) problem is chosen to stress the performance of the system.

Figure 5 shows the mean value of the cost $L(\omega)$ of the sampled paths with respect to the number of tours completed by ants, averaged over 26 runs. Error bars indicate 95% confidence intervals. Parameter settings are similar to those used in [4]. In this case, applying 'SEoF plain' significantly reduces the improvement obtained by using elite selection in terms of convergence speed. Applying 'SEoF entropy', the performance of the system is much less impaired.

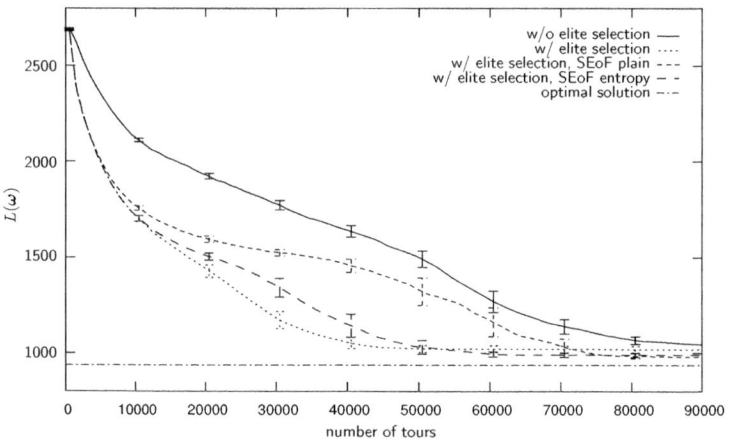

Fig. 5. 26 node TSP example

Appendix B: Fast Adaptation on Reheating Changes

In the simple illustrative example used throughout this paper, the performance of CEAS on reheating changes is worse with elite selection than without, and the performance of the system without elite selection is used as a reference to define stagnation. When the set of candidate solutions is larger, applying elite selection

---

[7] http://www.iwr.uni-heidelberg.de/groups/comopt/software/TSPLIB95

generally improves the performance of the system also on reheating changes by focusing the search around the best solutions. However, considering stagnation in a broader sense, the causes of stagnation listed in Section 4 still apply and prevent the system from fast adaptation on reheating changes. Now, SEoF is not designed to match the performance of the system without elite selection and it improves the adaptivity of the system in general. This is demonstrated by applying CEAS to find and maintain the minimum cost path between nodes 0 and 9 in the 10 node network depicted in Figure 6.

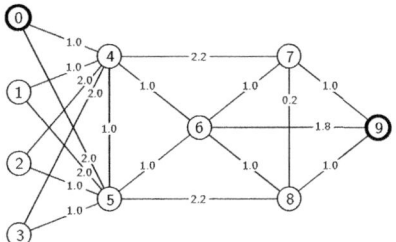

Fig. 6. 10 node network

The cost $L(\omega_t)$ of a path is given by the sum of the delays of each link. Parameters are set as in Section 4. At $t = 0$ [s], all the links have the delay values (in ms) given in Figure 6 and the best path is $\langle 0, 4, 6, 9 \rangle$ (3.8 [ms]). At $t = 10000$ [s], the cost of the link between nodes 6 and 9 is increased to 2.2 [ms] (reheating change) and there are then two best paths, $\langle 0, 4, 6, 7, 9 \rangle$ and $\langle 0, 4, 6, 8, 9 \rangle$ (4.0 [ms]). Figure 7 shows the mean cost of the paths sampled by normal ants averaged over 30 replications. Error bars indicate 95% confidence intervals. Applying SEoF, the time t_r to converge after the cost increase is significantly reduced.

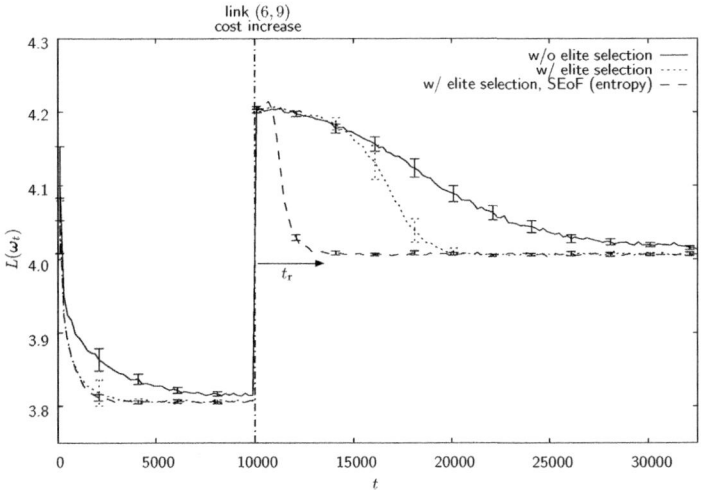

Fig. 7. Average cost of the paths sampled by normal ants

iNet-EGT: An Evolutionarily Stable Adaptation Framework for Network Applications

Chonho Lee[1], Junichi Suzuki[1], and Athanasios V. Vasilakos[2]

[1] Department of Computer Science
University of Massachusetts, Boston, USA
{chonho,jxs}@cs.umb.edu
[2] Department of Computer and Telecommunication Engineering
University of Western Macedonia, Greece
vasilako@ath.forthnet.gr

Abstract. This paper studies a bio-inspired framework, iNet-EGT, to build autonomous adaptive network applications. In iNet-EGT, each application is designed as a set of agents, each of which provides a functional service and possesses biological behaviors such as migration, replication and death. iNet-EGT implements an adaptive behavior selection mechanism for agents. It is designed after an immune process that produces specific antibodies to antigens (e.g., viruses) for eliminating them. iNet-EGT models a set of network conditions (e.g., workload and resource availability) as an antigen and an agent behavior as an antibody. iNet-EGT allows each agent to autonomously sense its surrounding network conditions (an antigen) and select a behavior (an antibody) according to the conditions. This behavior selection process is modeled as a series of evolutionary games among behaviors. It is theoretically proved to converge to an evolutionarily stable (ES) equilibrium; a specific (i.e., ES) behavior is always selected as the most rational behavior against a particular set of network conditions. This means that iNet-EGT allows every agent to always perform behaviors in a rational and adaptive manner. Simulation results verify this; agents invoke rational (i.e., ES) behaviors and adapt their performance to dynamic network conditions.

Keywords: Artificial immune systems, Evolutionary game theory, Biologically-inspired networking, Autonomous and adaptive networks.

1 Introduction

Network applications face critical challenges such as *autonomy*–the ability to operate with minimal human intervention and *adaptability*–the ability to adjust their operations to dynamic changes in network conditions such as workload and resource availability. In order to address these challenges, this paper investigates a biologically-inspired framework to design autonomous adaptive network applications. Based on an observation that various biological systems (e.g., bee colonies) have successfully attained autonomy and adaptability, the authors of

E. Altman et al. (Eds.): Bionetics 2009, LNICST 39, pp. 36–49, 2010.

the paper believe that, if network applications are designed after key biological mechanisms, they may be able to attain autonomy and adaptability as well.

In this paper, each network application is designed as a decentralized group of software agents. This is analogous to a bee colony (an application) consisting of multiple bees (agents). Each agent implements a functional service and follows biological behaviors such as migration, replication and death. This paper focuses on an adaptive behavior selection mechanism for agents. The proposed mechanism, called iNet-EGT, is designed after immunological *antigen-antibody reaction*, which produces antibodies specific to antigens (e.g., viruses) for eliminating them. iNet-EGT models a set of network conditions (e.g., workload and resource availability) as an antigen and an agent behavior as an antibody. Each agent contains iNet-EGT as its behavior selection mechanism. iNet-EGT allows each agent to autonomously sense its surrounding network conditions (an antigen) and select a behavior (an antibody) suitable for the sensed conditions. For example, agents may invoke the replication behavior at the network hosts that accept a large number of user requests for their services. This leads to the adaptation of agent availability; agents can improve their throughput.

In iNet-EGT, antigen-antibody reaction (i.e., behavior selection) process is modeled with evolutionary game theory. Each agent contains a set (or population) of behaviors. In a behavior selection process, randomly-selected two behaviors play a game. Each game distinguishes a winning and a losing behavior according to their payoff values computed based on the current network conditions. The winner replicates itself and increases its share in the population. The loser disappears in the population. Through multiple games performed repeatedly in the population the population state (behavior distribution) changes. Through theoretical analysis, iNet-EGT guarantees that the population state converges to an equilibrium where the population is occupied by only one type of behaviors, called strictly dominant behaviors. Each agent invokes a strictly dominant behavior as the most rational behavior against the current network conditions.

iNet-EGT theoretically proves that the population state is evolutionarily stable (ES) when it is on an equilibrium. An ES state is the state that, regardless of the initial population state, the population state always converges to. In this state, no other behaviors except strictly dominant one can dominate the population. Given this property, iNet-EGT guarantees that all agents deterministically invoke a specific ES behavior under a particular set of network conditions. Simulation results verify this theoretical analysis; agents seek equilibria to invoke ES behaviors and adapt their performance to dynamic network conditions.

2   Backgroud: Evolutionary Game Theory

Game theory studies strategic selection of behaviors in interactions among rational players. In a game, given a set of strategies, each player strives to find a strategy that optimizes its own payoff depending on the others' strategy choices. Game theory seeks such strategies for all rational players as a solution, called

Nash equilibrium (NE), where no players can gain extra payoff by unilaterally changing his strategy.

Evolutionary game theory (EGT) is an application of game theory to biological contexts to analyze population dynamics and stability in biological systems. In EGT, games are played repeatedly by players randomly drawn from the population [1, 2]. In general, EGT considers two major evolutionary mechanisms: *mutation*, which injects varieties on genes, and *selection*, which favors some varieties over others based on their fitness to the environment. Mutation is considered in the notion of evolutionarily stable strategies (ESS), which is a refinement of NE. Selection is considered in the replicator dynamics (RD) model.

2.1   Evolutionarily Stable Strategies

ESS is a key concept in EGT. A population following such a strategy is invincible. Specifically, suppose that the initial population is programmed to play a certain pure or mixed strategy x (the incumbent strategy). Then, let a small population share of players $\epsilon \in (0, 1)$ play a different pure or mixed strategy y (the mutant strategy). Hence, if a player is drawn to play the game, the probabilities that its opponent plays the incumbent strategy x and the mutant strategy y are $1 - \epsilon$ and ϵ, respectively. The player's payoff of such a game is the same as that of a game where the player plays the mixed strategy $w = \epsilon y + (1 - \epsilon)x$. The payoffs of players with strategies x and y given that the opponent adopts strategy w are denoted by $U(x, w)$ and $U(y, w)$, respectively.

Definition 1. *A strategy x is called evolutionarily stable if, for every strategy $y \neq x$, a certain $\bar{\epsilon} \in (0, 1)$ exists, such that the inequality*

$$U(x, \epsilon y + (1 - \epsilon)x) > U(y, \epsilon y + (1 - \epsilon)x) \tag{1}$$

holds for all $\epsilon \in (0, \bar{\epsilon})$.

In the special case where the payoff function is linear, $U(x, w)$ and $U(y, w)$ can be written as the expected payoffs for players with strategies x and y, and Equation (1) yields

$$(1 - \epsilon)U(x, x) + \epsilon U(x, y) > (1 - \epsilon)U(y, x) + \epsilon U(y, y) \tag{2}$$

If ϵ is close to zero, Equation (2) yields either

$$U(x, x) > U(y, x), \quad \text{or} \tag{3}$$
$$U(x, x) = U(y, x) \quad \text{and} \quad U(x, y) > U(y, y) \tag{4}$$

Hence, it becomes obvious that an ESS must be a NE; otherwise, Equation (3) or (4) do not hold.

2.2   Replicator Dynamics

The replicator dynamics, first proposed by Taylor and Jonker [4], specifies how population shares associated with different pure strategies evolve over time. In replicator dynamics players are programmed to play only pure strategies. To define the replicator dynamics, consider a large but finite population of players programmed to play pure strategy $k \in K$, where K is the set of strategies. At any instant t, let $\lambda_k(t) \geq 0$ be the number of players programmed to play pure strategy k. The total population of players is given by $\lambda(t) = \sum_{k \in K} \lambda_k(t)$. Let $x_k(t) = \lambda_k(t)/\lambda(t)$ be the fraction of players using pure strategy k at time t. The associated population state is defined by the vector $\mathbf{x}(t) = [x_1(t), \cdots, x_k(t), \cdots, x_K(t)]$. Then, the expected payoff of using pure strategy k given that the population is in state $\mathbf{x}$ is $U(k, \mathbf{x})$ and the *population average payoff*, that is the payoff of a player drawn randomly from the population, is $U(\mathbf{x}, \mathbf{x}) = \sum_{k=1}^{K} x_k \cdot U(k, \mathbf{x})$. Suppose that payoffs are proportional to the reproduction rate of each player and, furthermore, that a strategy profile is inherited. This leads to the following dynamics for the population shares x_k

$$\dot{x}_k = x_k \cdot [U(k, \mathbf{x}) - U(\mathbf{x}, \mathbf{x})] \tag{5}$$

where x_k is the time derivative of x_k. The equation states that populations with better (worse) strategies than average grow (shrink). However, there are cases when even *a strictly dominated strategy* may gain more than average. Hence, it is not a priori clear whether if such strategies get wiped out in the replicator dynamics. The following theorem answers this question [1]:

Theorem 1. *If a pure strategy k is strictly dominated then $\xi_k(t, x^0)_{t \to \infty} \to 0$, where $\xi_k(t, x^0)$ is the population at time t and x^0 is the initial state.*

On the other hand, it should be noted that the ratio x_k/x_ℓ of two population shares $x_k > 0$ and $x_\ell > 0$ increases with time if the strictly dominated strategy k gains a higher payoff than the strictly dominated strategy ℓ. This is a direct result of Equation (5) and may be expressed analytically via

$$\frac{d}{dt} \left[\frac{x_k}{x_\ell} \right] = [U(k, \mathbf{x}) - U(\ell, \mathbf{x})] \frac{x_k}{x_\ell} \tag{6}$$

From Equation (6), it is evident that even suboptimal strategies could temporarily increase their share before being wiped out in the long run. However, there is a close connection between NE and the steady states of the replicator dynamics, which is states where the population shares do not change their strategies over time. Thus, since in NE all strategies have the same average payoff, every NE is a steady state. The reverse is not always true: Steady states are not necessarily NE, e.g., any state where all players use the same pure strategy is a steady state, but, it is not stable [1].

   In this paper, a single fixed-sized population model is used; also, discrete time (i.e., generational) model is assumed.

3 iNet-EGT

This section describes how iNet-EGT is designed after an immunological process.

3.1 The Natural Immune System

The immune system is an adaptive defense mechanism to regulate the body against dynamic environmental changes such as antigen invasions. Through a number of interactions among various white blood cells (e.g., macrophages and lymphocytes) and molecules (e.g., antibodies), the immune system evokes *antigen-antibody reaction* to produce antibodies specific to detected antigens.

Antibodies form a network and communicate with each other [5]. This immune network is formed with stimulation and suppression relationships among antibodies. An antibody stimulates or suppresses another one based on its affinity to an antigen. Through the stimulation and suppression relationships, antibodies dynamically change their population. For example, a stimulated/suppressed antibody replicates/dies and increases/decreases its population. The population of specific antibodies rapidly increases following the recognition of an antigen and decreases again after eliminating the antigen. Through this self-regulation mechanism, adaptive immune response is an emergent product of interactions among antibodies.

3.2 Immunologically-Inspired Adaptation Behavior Selection

An agent contains iNet-EGT as its own immune system. iNet-EGT implements an adaptive behavior selection mechanism for an agent by following antigen-antibody reaction in the natural immune system. It is designed to allow each agent to autonomously sense a set of its surrounding network conditions (an antigen) and adaptively perform a behavior (an antibody) suitable for the conditions (Figure 1).

In iNet-EGT, an antigen consists of network conditions: $C = \{c_1, c_2, \cdots, c_L\}$ where L denotes the number of network conditions that each agent senses. For example, $C = \{100 : Workload, 35 : ResourceUtilization\}$ may mean 100 user requests per minute as workload and 35% memory utilization as resource utilization. Each antibody represents one of behavior types (e.g., migration, replication and death): $B = \{b_1, b_2, \cdots, b_M\}$ where M denotes the number of behavior types that each agent invokes.

In iNet-EGT, behavior selection (i.e., antigen-antibody reaction) is modeled based on evolutionary game theory. iNet-EGT executes an evolutionary game in a population of behaviors (antibodies) and determines one of the behaviors to be invoked by an agent. After initializing the population, randomly-selected two behaviors repeatedly play games in the population. Each game distinguishes a winning behavior and a losing behavior according to their fitness (or payoff) values that are computed based on the current network conditions. The loser disappears in the population. The winner replicates itself and increases its share in the population. The winner is also mutated at a certain probability in order to

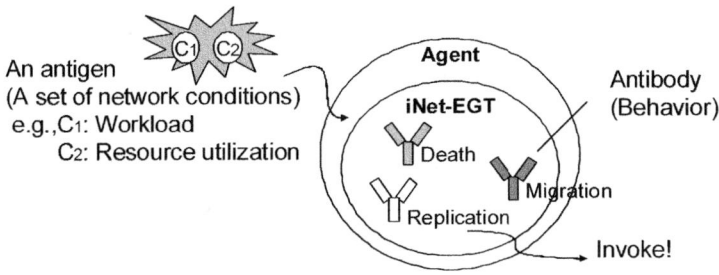

Fig. 1. Antigen-antibody reaction

react to future changes in network conditions. Then, eventually one behavior in a population is selected as the behavior invoked by an agent. Figure 2 presents a pseudocode of the implemented behavior selection process.

```
BehaviorSelection()
// P: Population, W: A set of winners, M: A set of the mutated
// a_b: A behavior invoked by an agent
main
INITIALIZEPOPULATION(P)
while (the termination condition is not satisfied)
      ⎰ W, M ← Φ, n ← P/2
      ⎢ for i ← 0 to n
      ⎢        ⎰ COMPUTEFITNESSVALUE(P)
      ⎢        ⎢ {behavior_1, behavior_2} ← SELECT(P)
  do ⎨    do ⎨ P ← P − {behavior_1, behavior_2}
      ⎢        ⎢ winner ← PERFORMGAME(behavior_1, behavior_2)
      ⎢        ⎢ W ← W ∪ winner
      ⎢        ⎱ M ← M ∪ MUTATE(winner)
      ⎱ P ← W ∪ M
a_b ← {b | b ∈ P, X_b(t) > th}
```

Fig. 2. Pseudocode of behavior selection in iNet-EGT

Figure 3 describes how behavior selection process works in each generation. A population is implemented as an array of behaviors, each behavior is associated with one of behavior types (i.e., actual actions as strategies) such as migration and replication. A population state at time t represents behavior distribution in the population, and it is denoted by $X(t) = \{x_1(t), x_2(t), \cdots, x_M(t)\}$ where $x_b(t)$ is the population share of a behavior type b, i.e., $x_b = \frac{n_b}{N}$, where $N = \sum_{b \in B} n_b$ where n_b is the number of behaviors with a behavior type b; so $\sum_{b \in B} x_b = 1$. Initially, behavior types are evenly distributed into behaviors in a population. For example, if the size of population is 100 and the size of behavior set is 4, then each 25 behaviors has one of the behavior types, i.e., $X(0) = \{.25, .25, .25, .25\}$.

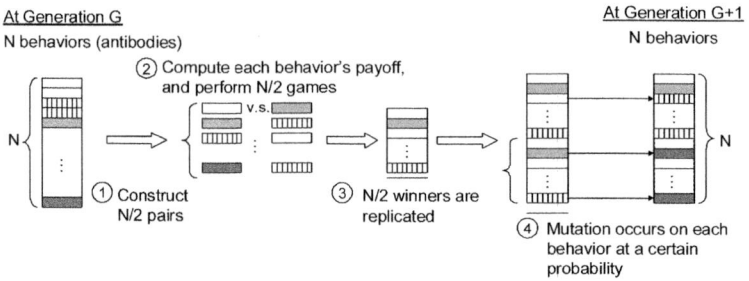

Fig. 3. Behavior Selection in each Generation

Each behavior (antibody) in a population computes its own fitness value (as affinity) against a set of network conditions (antigen). F_b denotes a fitness value of behavior b. In the current study, 4 behaviors (M: migration, R: replication, D: death, N: do-nothing) and 2 network conditions $C = \{c_1, c_2\}$ are considered. F is designed as follows:

$$F_{M^k} = c_1^k + c_2^k, \qquad F_D = (1 - c_1) - c_2$$
$$F_R = c_1 + c_2, \qquad F_N = (1 - c_1) + (1 - c_2)$$

Two network conditions, *Queue length* (c_1) and *Request rate* (c_2), indicate the spatial and temporal changes of the network environment in terms of workload. Queue length c_1^k is the number of user requests waiting to be processed in a queue at node k. Request rate c_2^k is the difference between the number of user requests received for a particular time period and that for the previous time period. Assume that a node maintains the number of user requests, $R(\Delta T)$, for a time period between $t - 1$ and t. Request rate is computed as $c_2(t) = R(\Delta(T)) - R(\Delta(T - 1))$.

A game is performed between randomly paired behaviors. A behavior wins/loses against another one based on their fitness values. A losing behavior is removed from a population. A winning behavior survives for the next generation and makes its copy to increase its population share; in addition, the mutation occurs on each copied behavior at a certain probability to change its behavior to another. iNet-EGT repeats the same process until the termination condition is satisfied. When one of the behaviors occupies the population based on the condition, $X_b(t) > th$, the behavior type b is selected. The threshold value th is set to 0.95 since the mutation probability is set to 0.05 in simulation studies.

4   Stability Analysis

This section analyzes the stability of behavior selection in iNet-EGT by showing that a population state converges to an evolutionarily stable state (or an asymptotically stable state) in three steps: (1) The dynamics of population state change over time is formalized as a set of differential equations, (2) The proposed behavior selection has equilibrium points, (3) The equilibrium points are asymptotically stable. First, in order to construct the differential equations, following terminologies and variables are defined.

- B denotes a set of behavior types. $B = \{b_1, b_2, \cdots, b_M\}$, and M denotes the number of behavior types.
- N denotes a population size. $N = \sum_{b \in B} n_b$ where n_b is the number of behaviors with a behavior type b.
- $X(t)$ denotes a population state at time t. $X(t) = \{x_1(t), x_2(t), \cdots, x_M(t)\}$ where x_b is the population share of a behavior type b ($x_b = \frac{n_b}{N}; \sum_{b \in B} x_b = 1$).
- F_b is the fitness value of a behavior with a behavior type b.
- p_k^b denotes the probability that a behavior with a behavior type b is replicated by winning a game against the behavior with a behavior type k. It is computed by $p_k^b = x_b \cdot \phi(F_b - F_k)$ where $\phi(F_b - F_k)$ is the conditional probability that the fitness value of a behavior with a behavior type b is larger than that of a behavior type k.

How behaviors with a behavior type b change their population share is considered as the sum of difference between the number of behaviors which are replicated (win) and eliminated (lose) at a time; then it is formalized as follows (using a brevity $c_{bk} = \phi(F_b - F_k) - \phi(F_k - F_b)$).

$$\dot{x}_b = \sum_{k \in B, k \neq b} \{x_k p_k^b - x_b p_b^k\} = x_b \sum_{k \in B, k \neq b} x_k \{\phi(F_b - F_k) - \phi(F_k - F_b)\}$$

$$= x_b \sum_{k \in B, k \neq b} x_k \cdot c_{bk} \qquad (7)$$

Theorem 2. *If a behavior with a behavior type k is strictly dominated, then $x_k(t) \to 0$ as $t \to \infty$.*

In game theory, it is said that a strategy (behavior type) is strictly dominant if, regardless of what any other players (behaviors) select, a player with the strategy gains a strictly higher payoff than any others. If a behavior has a strictly dominant behavior type, than it is always better than any others in terms of a fitness value (payoff). It will increase its population share and occupy a population over time. So, if a behavior is strictly dominated, then the behavior disappear in a population over time.

Theorem 3. *The population state of an agent converges to an equilibrium.*

Proof. It is true that, according to the fitness function (Equation 7), behaviors with different behavior types have different fitness values under the same network conditions. In other words, under the particular network conditions, only one behavior has the highest fitness value among the others. Assume that $F_1 > F_2 > \cdots > F_M$, and by Theorem 1, a population state eventually converges to $X(t) = \{x_1(t), x_2(t), \cdots, x_M(t)\} = \{1, 0, \cdots, 0\}$ as an equilibrium. Differential equations should satisfy the constraint $\sum_{b \in B} x_b = 1$. $\qquad \square$

Theorem 4. *The equilibrium of behavior selection in iNet-EGT is evolutionarily stable (i.e., asymptotically stable).*

Proof. At the equilibrium where $X = \{1, 0, \cdots, 0\}$, a set of differential equations can be rewritten in the downsized by substituting $x_1 = 1 - x_2 - \cdots - x_M$

$$\dot{z}_b = z_b[c_{b1}(1 - z_b) + \sum_{i=2, i \neq b}^{M} z_i \cdot c_{bi}] \quad \text{where} \quad b = 2,...,M \tag{8}$$

where $Z(t) = \{z_2(t), z_3(t), \cdots, z_M(t)\}$ denotes the corresponding downsized population state, which is an equilibrium $Z_{eq} = \{0, 0, \cdots, 0\}$ of (M-1)-dimension based on Theorem 2.

To verify that a state at the equilibrium is an asymptotically stable state, show that all the Eigenvalues of Jaccobian matrix of the downsized population state has negative Real parts. The elements of Jaccobian matrix J are

$$J_{bk} = \left[\frac{\partial \dot{z}_b}{\partial z_k}\right]_{|Z=Z_{eq}} = \left[\frac{\partial z_b[c_{b1}(1 - z_b) + \sum_{i=2, i \neq b}^{M} z_i \cdot c_{bi}]}{\partial z_k}\right]_{|Z=Z_{eq}} \tag{9}$$

$$where \ b, k = 2, ..., M$$

Therefore, Jaccobian matrix J is given by

$$J = \begin{bmatrix} c_{21} & 0 & \cdots & 0 \\ 0 & c_{31} & \cdots & 0 \\ \vdots & \vdots & \ddots & \vdots \\ 0 & 0 & \cdots & c_{M1} \end{bmatrix} \tag{10}$$

where $c_{21}, c_{31}, \cdots, c_{M1}$ are the Eigenvalues of J. According to Theorem 2, $c_{b1} = -\phi(F_1 - F_b) < 0$ for every b; therefore, $Z_{eq} = \{0, 0, \cdots, 0\}$ is asymptotically stable. An agent deterministically invokes a specific behavior (i.e., ES behavior) under a particular set of network conditions.   □

5   Simulation Results

This section evaluates iNet-EGT through simulations. Figure 4 shows a simulated network, which is a server farm consisting of 16 (4 x 4) hosts in a grid topology. User requests travel from users to agents via user access point. This simulation study assumes that a single (emulated) user runs on the access point and sends user requests to agents.

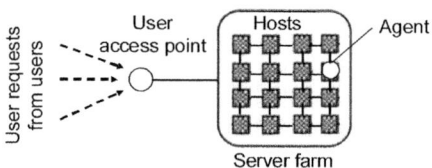

Fig. 4. Simulated Network

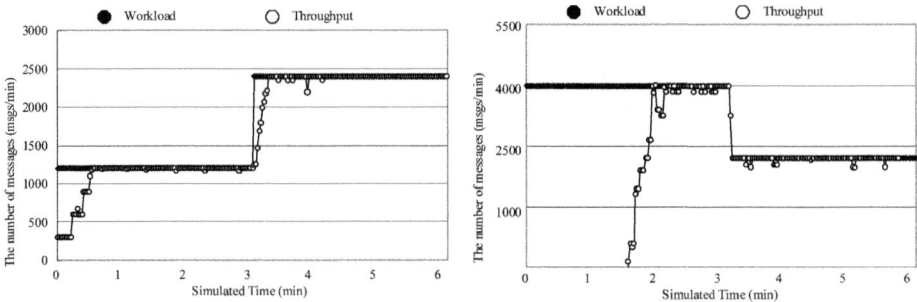

Fig. 5. Workload Type 1 and Throughput   **Fig. 6.** Workload Type 2 and Throughput

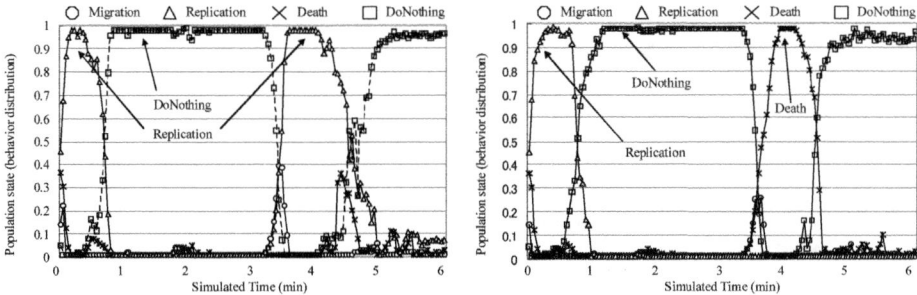

Fig. 7. Population State (Workload #1)   **Fig. 8.** Population State (Workload #2)

At the beginning of a simulation, an agent is deployed on a randomly-selected host in the network. Each agent has its own iNet-EGT that contains a population of 100 behaviors. (25 behaviors are of each of four behavior types: migration, replication, death and do-nothing). Mutation rate and behavior selection threshold are set to 0.1 and 0.95, respectively. Figures 5 and 6 show two different types of changes in workload (i.e., the number of user requests) given to agents.

Figure 7 shows how population state (behavior distribution) changes over time in an agent deployed at the beginning of a simulation. (The two figures show the changes in population state against the workload type 1 and 2, respectively.) In Figure 7, the number of replication behaviors increases in the first 15 seconds, and population state converges to an ES state. Then, the do-nothing behavior takes over the replication behavior to dominate the population; the population converges to another ES state. The second ES state emerges because agents finish adapting their availability with the replication behavior in the first one minute to efficiently process incoming user requests. (See Figure 9 for the changes in agent availability under the workload type 1.) This ES state continues until workload spikes at the third minute. Upon the workload spike, the replication behavior dominates the behavior population again. Once agent availability adapts to the workload spike, the do-nothing behavior takes over the replication behavior.

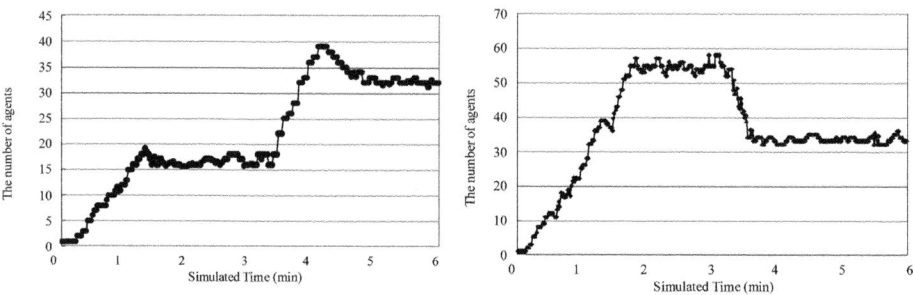

Fig. 9. Agent Availability (Workload #1) **Fig. 10.** Agent Availability (Workload #2)

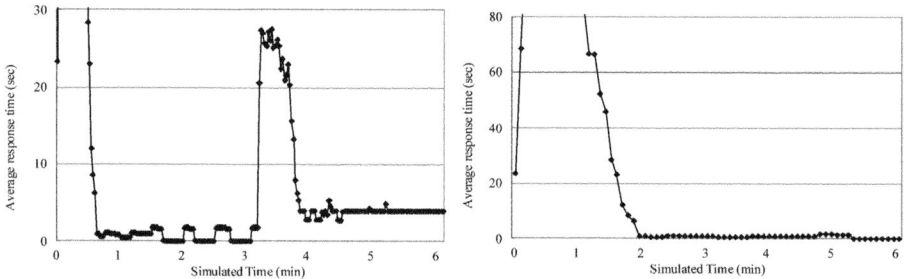

Fig. 11. Response Time (Workload #1)   **Fig. 12.** Response Time (Workload #2)

Figure 8 shows the changes in population state under the workload type 2. The changes are similar to those in Figure 7 except that the death behavior dominates the behavior population when workload drops. See Figure 10 for the changes in agent availability under the workload type 2. As shown in Figures 7 and 8 , iNet-EGT allows agents to successfully seek ES equilibria in their behavior selection according to dynamic network conditions.

Figures 9 and 10 show how agent availability (i.e., the number of agents) changes over time when the workload type 1 and 2 are given to agents, respectively. The two figures demonstrate that agents adapt their availability by invoking behaviors according to the ES states they are on. (See also Figures 7 and 8.)

Figures 5 and 11 show the throughput (i.e., the number of processed requests per minute) and response time that agents yield for users under the workload type 1. Figures 6 and 12 show the throughput and response time results under the workload type 2. At the beginning of a simulation, only one agent is deployed; it cannot efficiently process all user requests. As a result, throughput is low, and response time is high. However, as agents performs their behaviors by seeking ES states (Figures 7 and 8), they adapt their throughput and response time to dynamic network conditions.

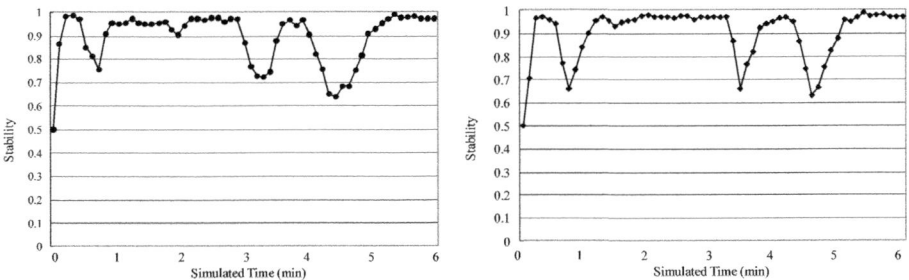

Fig. 13. Stability (Workload #1)   **Fig. 14.** Stability (Workload #2)

Figure 13 and 14 show the average stability of the behavior populations that agents possess. It is measured as follows:

$$S_{avg}(t) = \frac{1}{A(t)} \sum_i \max_{b \in B}\{x_b(t)\} \tag{11}$$

where $A(t)$ denotes the total number of agents. i indexes agents. b indexes behavior types $(1,\cdots,4)$. Agents seek equilibria to invoke evolutionarily stable behaviors. For example, when agents sufficiently adapt their availability to the workload at around 0:30, $S_{avg}(t)$ decreases because the number of replication and do-nothing behaviors change. However, soon or later agents increase the number of do-nothing behaviors, and a population state converges to the stable state again. In addition, the likelihood of agents operating at a stable state during a simulation run (e.g., 6 min) is observed as *stability*. It is measured as how long agents operate at equilibria during the simulation run (i.e., [time(sec) for $S_{avg}(t) > 0.95]/[6$ min=360 sec]). Along the workload type 1, the stability is about 82%. For the workload type 2, the stability is about 86%.

6   Related Work

iNet-EGT is an extension to its predecessor called iNet [6]. iNet-EGT and iNet share the same goal; immunologically-inspired adaptive behavior selection for agents. However, they are different in their approaches to design antigen-antibody reaction and antibody evolution. iNet designs antigen-antibody reaction based on a model built with the immune network hypothesis [5] and designs antibody evolution with a genetic algorithm. iNet-EGT takes evolutionary game theoretic approach to design antigen-antibody reaction and antibody evolution. It guarantees stability in behavior selection while iNet does not. iNet-EGT is the first attempt to model an artificial immune system based on EGT.

Conventional game theory has been introduced to several aspects in network systems; e.g., job allocation [7], security [8, 9, 10] and routing [11]. They focus on the rationality of behavior selection in static network environments; however, they do not consider adaptation in dynamic network environments. [12, 13] leverage EGT to formulate rational and adaptive routing decisions to dynamic network environments. Unlike [12, 13], iNet-EGT performs the mutation operation in the behavior selection to better adapt to future changes in network environments.

[14, 15, 16, 17] study adaptive behavior selection mechanisms for agent-based systems. [14] proposes a rule-based mechanism, which is similar to iNet-EGT in that it implements deterministic behavior selection. However, unlike [14], iNet-EGT guarantees stability in behavior selection. [15, 16, 17] consider non-deterministic behavior selection with stochastic algorithms. In contrast, iNet-EGT considers determinism in behavior selection to guarantee its stability.

7 Conclusion

This paper proposes and evaluates a bio-inspired framework, iNet-EGT, which aids building autonomous and adaptive network applications. iNet-EGT is designed after antigen-antibody reaction in the immune system. The reaction process is modeled as a series of evolutionary games among behaviors. It is theoretically proved to converge to an evolutionarily stable (ES) equilibrium. This means that iNet-EGT allows every agent to always perform behaviors in a rational and adaptive manner. Simulation results verify this; agents invoke rational (i.e., ES) behaviors and adapt their performance to dynamic network conditions.

References

1. Weibull, J.W.: Evolutionary Game Theory. MIT Press, Cambridge (1996)
2. Nowak, M.A.: Evolutionary Dynamics: Exploring the Equations of Life. Harvard University Press (2006)
3. Fudenberg, D., Levin, D.K.: The theory of learning in games. MIT Press, Cambridge (1998)
4. Taylor, P., Jonker, L.: Mathematical Biosciences, 16
5. Jerne, N.K.: Idiotypic networks and other preconceived ideas. Immunological Review (1984)
6. Lee, C., Wada, H., Suzuki, J.: Towards a biologically-inspired architecture for self-regulatory and evolvable network applications. In: Dressler, F., Carreras, I. (eds.) Advances in Biologically Inspired Information Systems. Springer, Heidelberg (2007)
7. Subrata, R., Zomaya, A.Y., Landfeldt, B.: Game theoretic approach for load balancing in computational grids. IEEE Transactions on Parallel and Distributed Systems 19(1) (2008)
8. Kodialam, M., Lakshman, T.V.: Detecting network intrusions via sampling: a game theoretic approach. In: Proc. of IEEE INFOCOM (April 2003)
9. Agah, A., Basu, K., Das, S.K.: Preventing dos attack in sensor networks: a game theoretic approach. In: Proc. of IEEE ICC (May 2005)
10. Otrok, H., Mehrandish, M., Assi, C., et al.: Game theoretic models for detecting network intrusions. Computer Communications 31 (June 2008)
11. Kannan, R., Iyengar, S.: Game theoretic models for reliable path-length and energy constrained routing with data aggregation in wireless sensor networks. IEEE J. on Selected Areas in Communications 22(6) (2004)

12. Vasilakos, A.V., Anastasopoulos, M.: Application of evolutionary game theory to wireless mesh networks. Studies in Comp. Intelligence. Springer, Heidelberg (2007)
13. Anastasopoulos, M.P., Petraki, D.K., Kannan, R., Vasilakos, A.V.: Tcp throughput adaptation in wimax networks using replicator dynamics. IEEE Transactions on Systems, Man, and Cybernetics (to appear)
14. Li, Z., Parashar, M.: Rudder: A rule-based multi-agent infrastructure for supporting autonomic grid applications. In: Proc. of IEEE ICAC (2004)
15. Wang, Y., Li, S., Chen, Q., Hu, W.: Biology inspired robot behavior selection mechanism: Using genetic algorithm. In: Proc. of LSMS (2007)
16. Damas, B.D., Custódio, L.: Emotion-based decision and learning using associative memory and statistical estimation. Informatica 27(2) (2003)
17. Kim, K.-J., Cho, S.-B.: Bn+bn: Behavior network with bayesian network for intelligent agent. In: Proc. of Australian Conf. on Artificial Intelligence (2003)

Situated Service Oriented Messaging for Opportunistic Networks

Juhani Latvakoski, Tomi Hautakoski, and Antti Iivari

VTT Technical Research Centre of Finland,
Kaitoväylä 1, P.O.Box 1100, FIN-90571 Oulu, Finland
{juhani.latvakoski,tomi.hautakoski,antti.iivari}@vtt.fi

Abstract. A novel concept for situated service oriented messaging applicable in the context of biologically inspired opportunistic networks has been provided in this paper. The solution utilizes different contextual information sources to create and update a view of the communicational situation. Smart diffusion of relevant control data between neighbouring nodes using novel swarm intelligence based method enables spreading of information only to the interested nodes without unnecessarily disturbing the non interested nodes. The evaluations done against epidemic routing protocol indicate that the proposed solution lowers the amount of transmissions in the network, thus reducing precious resource usage in the nodes. This is achieved without introducing further delays or deteriorations in the message delivery ratio.

Keywords: opportunistic communication, context awareness, service awareness.

1 Introduction

The number of wireless communicating embedded devices has continuously been increasing in recent years. Because of the inherent nature of such devices is to be both mobile and dynamic, it is obvious that the destination of communication is not necessarily reachable at the time of communication need. This type of challenge has also been described previously in the context of InterPlaNetary networks (IPNs), Delay-Tolerant Networks (DTN) [1, 2, 3] and opportunistic networking [4]. A common essential feature for them is that the source and destination may never be connected to the same network at the same moment of time, but communication may be enabled on a hop-by-hop basis. In such a case, finding a route by means of Mobile Ad hoc Networks (MANETs) such as e.g. Ad hoc On-Demand Distance Vector (AODV) is not possible. To solve the problems, several proposals have been provided such as combination of DTN and MANET routing [5]. The problem of the referred *disconnected communication* is still rather open research item and it has been one starting point problem for this research.

The opportunistic routing can be categorized to e.g. dissemination based or context based routing [4, 6]. Dissemination based routing techniques aim to deliver messages to the destination by simply diffusing them all over the network. Usually the

E. Altman et al. (Eds.): Bionetics 2009, LNICST 39, pp. 50–64, 2010.

dissemination methods work by offering the messages to neighbor nodes, when they are in the radio coverage. The offering can consist of sending the full message data to the neighbors, who then apply various filtering techniques to lower the network load. Another approach is to send advertisements of the data available at the sender, and receiver can request for the data based on the advertisements. Finally the sender will respond with the actual data message. Examples of dissemination based routing techniques are Epidemic routing [7], Meeting and Visits protocol [8] and Network coding based routing protocol [9]. Dissemination based approach work quite well when contact opportunities are very common. However, the problem with the dissemination based routing may be the heavy load generated into the network, which may cause network congestion resource over-usage situations. The network traffic can be lowered by limiting allowed hops or number of copies of the messages. The context based routing applies information about the contextual situation to achieve more efficient routing. Examples of context based routing techniques are Context-Aware routing (CAR) [10] and MobySpace routing [11]. The context based approach can reduce the network load compared with dissemination based approaches. However, the reasoning of the next hop increases the needed amount of CPU and memory resources from the nodes.

As a contribution, the situated service oriented messaging concept, and the algorithms applicable in the context of biologically inspired networks has been provided. Simulations are applied to evaluate the usefulness of the concept. The contribution essentially differs from the dissemination and context aware routing, because here both situation and service awareness are applied to optimize the message forwarding. The contribution extends the situated message forwarding concept [12] in the sense that here a new set of algorithms, and a novel swarm-based service-oriented approach for efficient communication inside a network island have been provided.

The rest of this paper is organized as follows. Chapter 2 describes the situated service oriented delivery concept and related algorithms. Chapter 3 describes the simulation based evaluations of the provided methods. Finally, conclusions are provided in chapter 4.

2 Situated Service Oriented Messaging

2.1 Concept

It is assumed in this research that the service and situation awareness in the message forwarding will decrease the amount of the load in the network by decreasing the amount of useless traffic while still keeping the reliability of messaging high. In addition, it is assumed that the increase of self-organization capabilities in the system will enable better scalability. In our approach, the neighbourhood information is applied to increase situation awareness, and also service information is applied when deciding the delivery scope. The decision is made locally in each individual node according to principles of self-organization, and in such a way to enable better system scalability.

From the service and content point of view, the general requirement set for the spreading of information is that all the nodes, who are interested about the information, eventually receive the information content. For example, in the targeted advertising scenario, only a group of nodes should receive the information content, and the others are not interested about it at all. To save network resources the spreading of information should not be blind, but it should be *service aware* instead e.g. by enabling smart diffusion of relevant control data between neighbouring nodes. There should be no need to disturb the non interested nodes by delivering them information which provides little or no value to them. On the other hand, even non-interested nodes may act as carrier of information for a specific group.

The conceptual mode for the service and situated opportunistic communication is visualized in figure 1. The model consists of User nodes (U), tiny nodes (T) and access point (AP) nodes, according to the characteristics of the nodes. The T nodes are assumed to be small, limited capability nodes which cannot act as message forwarding nodes. The U nodes act as forwarding nodes, and the role of AP nodes is to route traffic from the opportunistic networks towards more static networks such as e.g. Internet. The referred nodes may belong to network islands, e.g. Bionets A, according to their communication ranges. There are three network islands visualized in figure 1.

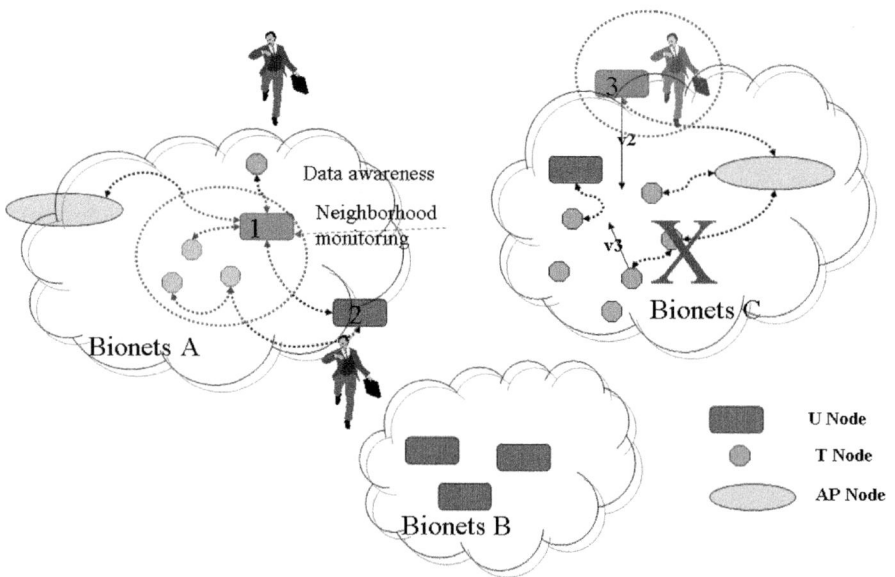

Fig. 1. Situated Service oriented opportunistic communication

The key features of the provided situated service oriented opportunistic communication concept are the following:

- Each U node monitors its neighborhood in order to collect real-time information about the situation in its' environment (neighborhood monitoring)
- The communication level in each U node receives information on the service content to enable smart data based message forwarding (data awareness)

- When deciding what to do for the incoming message, each U node operates according to the locally executed algorithm, which is used to decide whether the message should be forwarded or stored into the local memory.
- The solution consists of two components: Situated Adaptive Forwarding (SAF), and Service Oriented Forwarding (SOF).
- The algorithms are used as local reasoning engines for message forwarding to enable scalable opportunistic communication.

2.2 Situated Adaptive Forwarding (SAF)

SAF is based on monitoring statuses of nodes and their neighborhood, connectivity of nodes, their resource situation (CPU, message storage space) and classification of messages into classes. The solution consist of neighborhood discovery and message sending & message forwarding algorithms, each of which are described in the following.

2.2.1 Neighborhood Discovery Procedure

We begin by introducing the how neighboring nodes are found, and what information is exchanged among them. Every node has a Network Situation Database (NSDB) which contains context information relevant to forwarding of messages. An example entry of this database is shown in figure 2.

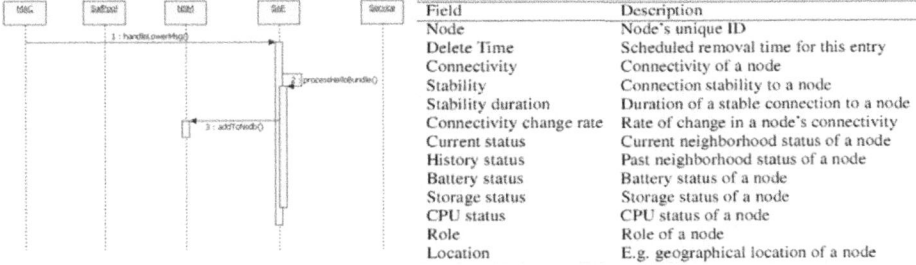

Field	Description
Node	Node's unique ID
Delete Time	Scheduled removal time for this entry
Connectivity	Connectivity of a node
Stability	Connection stability to a node
Stability duration	Duration of a stable connection to a node
Connectivity change rate	Rate of change in a node's connectivity
Current status	Current neighborhood status of a node
History status	Past neighborhood status of a node
Battery status	Battery status of a node
Storage status	Storage status of a node
CPU status	CPU status of a node
Role	Role of a node
Location	E.g. geographical location of a node

Fig. 2. A Network Situation Database (NSDB) entry and the NSDB refresh procedure

The discovery mechanism works by sending periodic HELLO messages. If a received HELLO came from an already known source node, its old entry in NSDB is refreshed with new values contained in the message (see figure 2).

When a new node is discovered (figure 3), a transfer is made if there are stored messages that should be sent also to the newly discovered node. Next, also those stored messages that are destined for any node in the newly met node's current network neighborhood, are forwarded. Finally, copies of such stored messages, which have Remaining Chance (RC) value bigger than 0.0 are sent to the discovered node. RC values are assigned for messages in the following way: If a message cannot be delivered immediately, because of the sender's current neighborhood or its' neighbors' total sum of chance for delivery doesn't reach the value assigned for the category of the message, it is also saved in the senders' message pool. Then, a RC value is given to a message based on its category, and delivery chances of those neighboring nodes that have gotten a copy of the message.

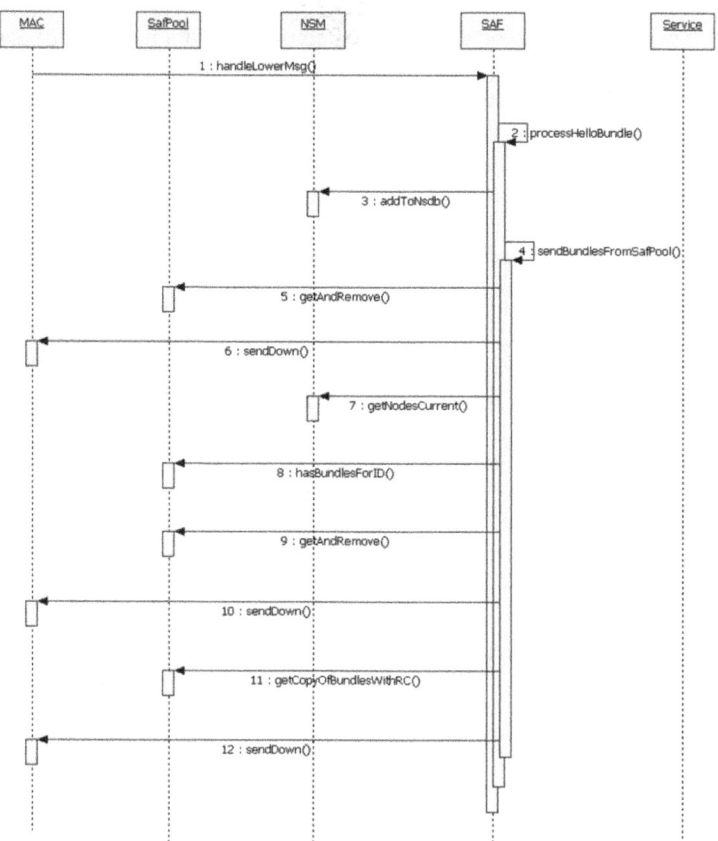

Fig. 3. A procedure when a new node is discovered

2.2.2 Message Sending and Forwarding Algorithms

To describe how the designed SAF model handles sending and reception of messages, sequence diagrams of different execution paths depending on the situation have been provided. Figures 4-6 provide explanations on how outgoing messages are handled. In figures 6-10, we show how incoming data messages are processed. The first three sequence diagrams all present the same principle case where the service layer needs to send a message to a certain node ID. At the service layer, a *sendBundle()* function is called which sends data using the following message structure: < | Type | Source | Destination | Sequence Number | Hop Count | Category | Role | Payload | >. The SAF module (at the network layer) receives the message and checks if the destination node ID is located in the vicinity of this node. In the case 1 (figure 4), this is expected to be true, and thus SAF creates a message representing a network layer message. This message is filled with headers and the service message as a payload. The headers include the following fields: < | Type | Source | Destination | Sequence Number | Purge Time | Category | Payload | >. The message is then sent to MAC module for transmission.

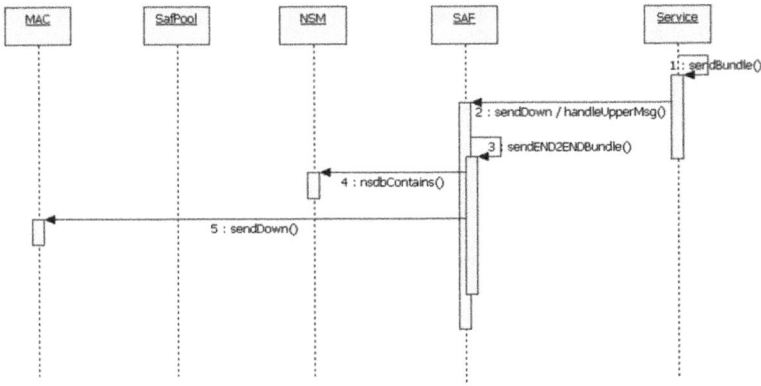

Fig. 4. Case 1, processing of an outgoing message

In case 2 (see figure 5), the destination ID is not a direct neighbor of the source node, and therefore it checks if the needed ID is within 2-hop range, which is true now and the message can be sent away.

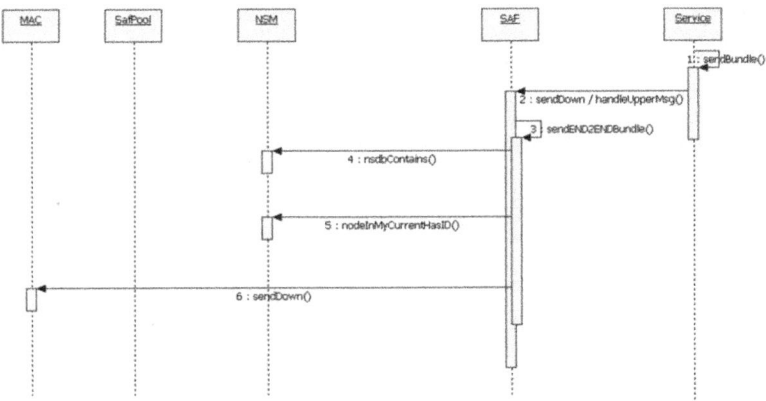

Fig. 5. Case 2, processing of an outgoing message

Case 3 (in figure 6) describes the most complex situation where the destination cannot be found within the current network neighborhood of the source node. Now dissemination of the message in the network in a best way possible given the situation at hand is started. A multi-keyed map consisting of key - value pairs is queried from the NSDB where each value is relative delivery probability and key indicates what neighboring node ID has that probability. That operation is followed by a iteration over the map to find out if the total sum of probability of our neighboring nodes exceeds required threshold for the processed message's category. Now, this is expected to be false, and more nodes are added to the map describing neighbors to which the

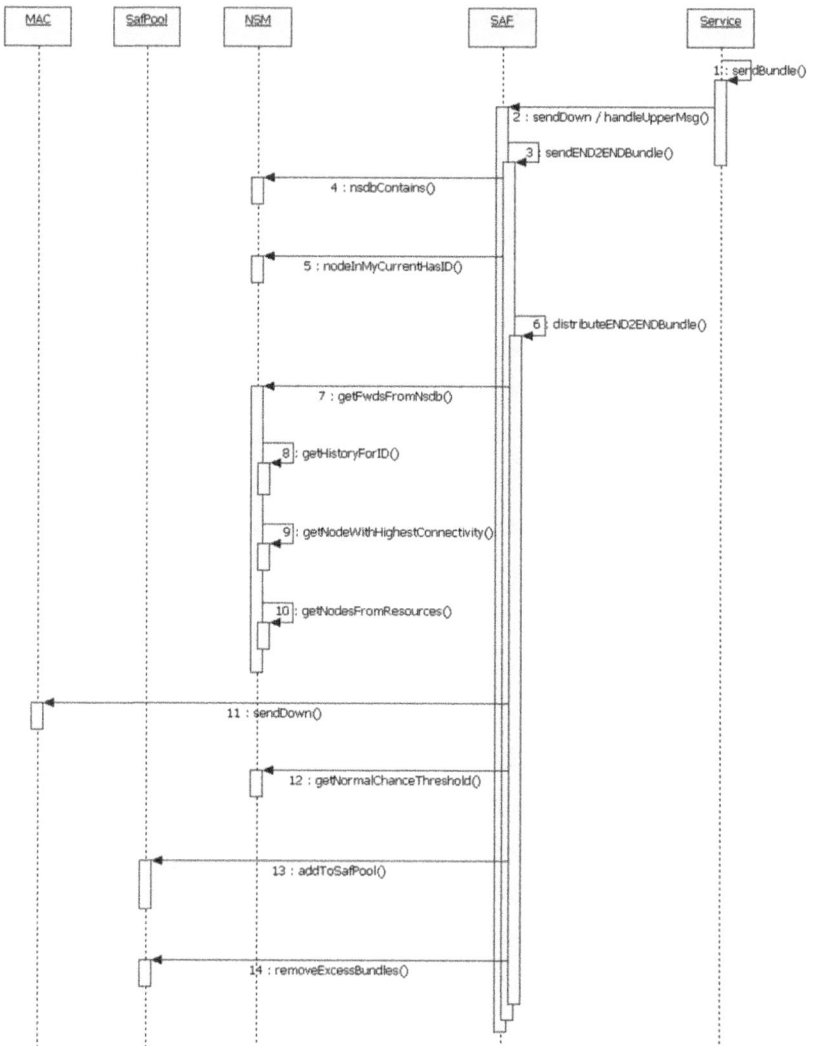

Fig. 6. Case 3, processing of an outgoing message

message is going to be sent. The delivery chance optionally increased by querying NSDB for the most connected node of the neighboring nodes as well as for nodes that have the most resources available, and adding those nodes to the map. Then the message is sent to all neighboring nodes listed in the map, and its RC value is updated as described in the earlier section.

In case 4 (figure 7) is where we begin to investigate how the SAF model handles incoming messages. First, messages are received from the MAC layer with *handleLowerMsg()* function, and in this case the incoming message was destined for this node; thus it is sent to service layer.

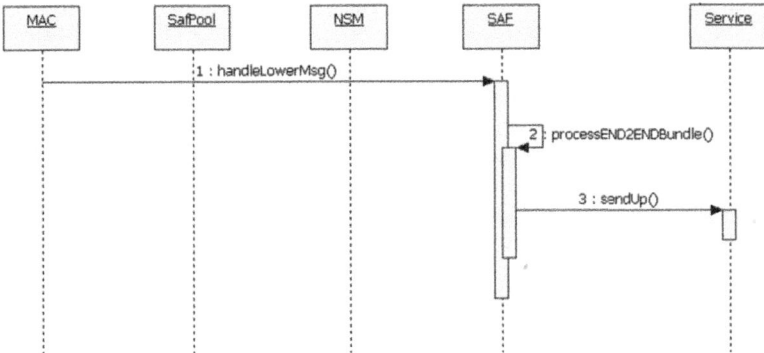

Fig. 7. Case 4, processing of an incoming message

Case 5 (see figure 8) describes a situation where the message was destined for some of the nodes that the receiver has in its neighborhood, and the message is forwarded without other procedures to the corresponding neighbor.

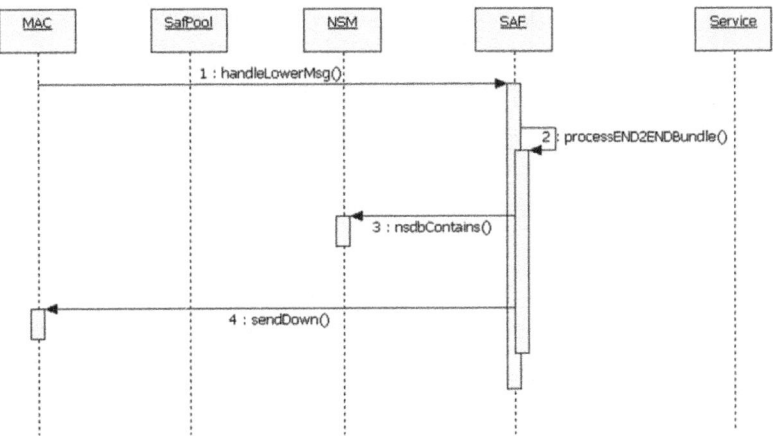

Fig. 8. Case 5, processing of an incoming message

We continue with case 6 (figure 9) where the first three steps are the same as before, but the destination ID of the message is searched within a 2-hop radius of the node's current neighborhood. A match is found and message is forwarded to a neighboring node which then forwards it again.

If none of the cases from 4 to 6 were applicable to an incoming message, case 7 (figure 10) includes a description of a sequence diagram that is used as a fail-safe option. The message is stored into the message pool of the sender. It is important to notice that in this case, the message is associated with a RC value of zero to limit too

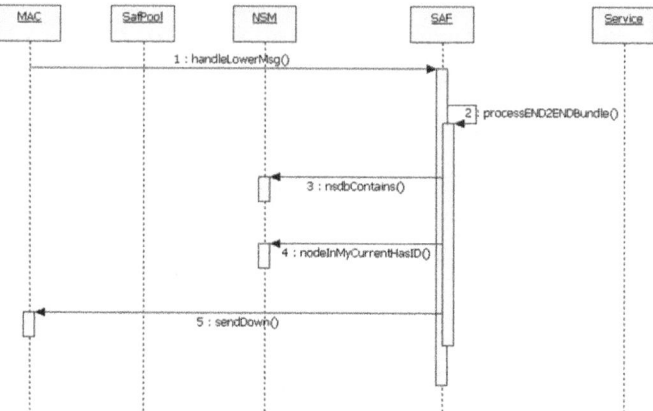

Fig. 9. Case 6, processing of an incoming message

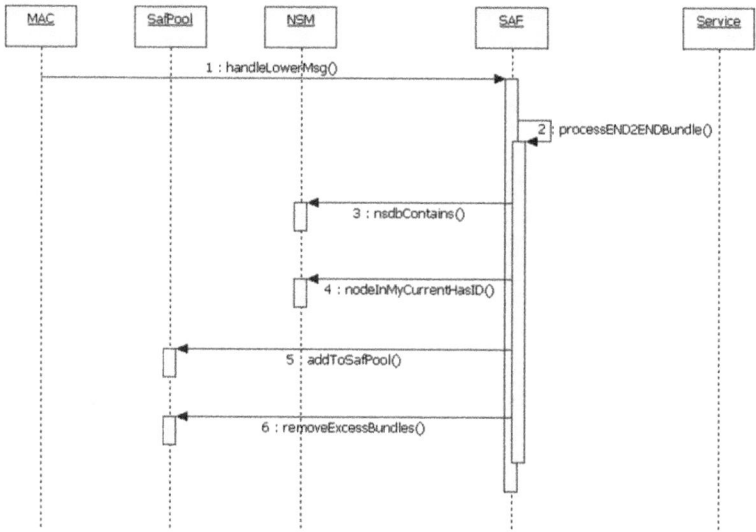

Fig. 10. Case 7, processing of an incoming message

resource exhausting dissemination. Finally, also the message pool is purged to have only valid messages and to keep it within given size limits.

2.3 Service Oriented Forwarding (SOF)

The service oriented message forwarding mechanism applies a swarm intelligence based approach, and its objective is to distribute messages inside a network island with minimal disturbance to non pertinent nodes in the network. In order to save network resources it is here proposed that the service oriented forwarding algorithm should not simply diffuse messages blindly to all the reachable nodes in the network. However, it ensures that the messages eventually reach a satisfactory amount of pertinent nodes even behind a zone of non-pertinent ones.

Since the situation will very likely change over time a smart and adaptive mechanism is required. The message diffusion algorithm needs to operate in a highly "distributed", self-organized manner and a single node will only need to follow a very simple set of rules. Swarm Intelligence (SI) based systems consist of (usually) unsophisticated agents interacting locally with each other and their environment that will eventually lead to the emergence of intelligent and coherent global behaviour. Taking into account the nature of the problem at hand and the characteristics of SI based systems discussed above, it is clear that SI will provide us an excellent basis on which it is possible to build algorithms that perform heavily distributed problem solving without centralized control or the utilization of a complicated global model. There are other communication network protocols based on SI, such as ANTNET. [13]. However, these algorithms mostly deal with problems associated with routing in standard IP networks and, as such, are not further discussed herein.

The Bio-inspired and SI based algorithm examined herein is inspired by the food foraging behaviour of a honey bee colony. This foraging process in a bee colony functions by deploying scout bees. These scout bees move randomly from one place to another in search of promising food sources. When the scout bees return to the hive, they communicate to the colony their findings. The information that they communicate contains, for example, the direction in which the food source is located and its fitness or "quality". Others have also used honey bees as inspiration for optimisation algorithms. For example, in [14] Pham *et al* describe a population-based search algorithm that mimics the food foraging behaviour of swarms of honey bees. By the application and slight adaptation of the principles discussed above, a highly adaptive decentralized networking technique is constructed that aims to efficiently disseminate messages to pertinent nodes. The source nodes will send scout messages at frequent intervals. Scout messages behave very much like scout bees in nature. The scout messages will randomly hop from node to node until they have reached a specific number of hops. They return to the originating node by going through the same path backwards. In addition to the node that originated the scout message, each node on the path of the scout message will extract and store information from the scout message. The nodes will store next hop and fitness values in a table relating to the service at hand. Over time, more and more of these scout messages will be sent and processed by the nodes in the network ensuring the emergence of applicable tables built using the information carried by scout messages at various nodes. Due to the high degree of randomness and the decentralized nature of the mechanism, the contents of these tables will keep adapting to reflect the changes in the network environment. In the beginning stage, a node initiates the process by generating and sending a scout message randomly to one neighbour. Then the neighbouring node stores information about the originating node, updates the information in the scout message and sends it to the next random neighbour (though, not a node that has already forwarded this particular message). This process continues, with each intermediate node gaining new information about the nodes on the scouts path, until a "dead-end" or the attainment of maximum hops specified for this scout message. The final node on the scout messages path will store and update information as before, but instead of sending it to a new random neighbour node, it sends it back to the previous node on the scout messages path. All intermediate nodes will now gain information also from the "forward" direction of the scout messages path while one by one returning the scout message back to the originating node through the same path, as shown in figure 11.

```
Algorithm 1 The process for handling a received scout message.
UPDATESCOUTMEMORY()                  ▷ extract and store information from the message
if scout.sourceAddress == myAddress then          ▷ if this is the originating node
    delete scout;
else if scout.isReturning == true  then       ▷ if the scout message is on its way back home
    RETURNSCOUT()                         ▷ send the scout message back to the source
else            ▷ otherwise the scout message must be forwarded to the next random neighbor
    scout.path ←myAddress              ▷ insert this nodes address into the scouts path
    if scout.serviceID == myServiceID then        ▷ increment the fitness count if necessary
        scout.fitness + 1
    end if
    if scout.hopCount == scout.maxHopCount then       ▷ if maxhops has been reached
        RETURNSCOUT()                       ▷ send the scout message back to the source
        return;
    end if
    GETRANDOMNEXTHOP()
    FORWARDSCOUT()                       ▷ forward the scout to a new random node
end if
```

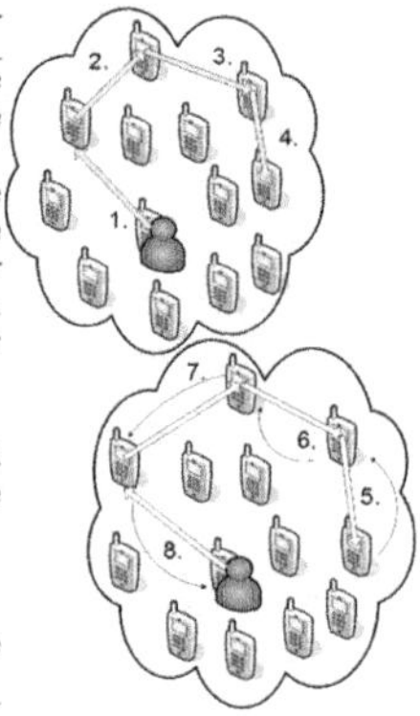

Fig. 11. Handling of a received scout message and the random hopping nature of the scout messages

Due to this ongoing process, the knowledge of the service situation in the neighbourhood is distributed among the nodes in the network island i.e. to which direction messages pertaining to a specific service should be forwarded. The longer this process goes on, the more such knowledge will be gained by the nodes receiving and forwarding scout messages.

3   Evaluation Results

3.1 Simulations

For evaluating feasibility of the SAF model, it was implemented in the OMNeT++ simulator against the popular Epidemic Routing model. The key modules, that are present in the sequence diagrams (figures 2-10), are: 1) "SAF" which implements the actual forwarding at network layer, 2) "NSM" for keeping the NSDB, 3) "SafPool" for storing messages and 4) "CSM" for having a more abstract view of different contexts (e.g. resources) that affect the communicational situation. Figure 12 lists used simulation parameters in the evaluations. Mobility pattern which the nodes followed was A Modified Reference Point Group Mobility Model with Dynamic Clustering (MRPGDC) described in [15]. Its models dynamic movement of human groups with a possibility that a person can leave his/hers group and start to follow a new group of people. For example, in real life these kinds of situations can be observed when people move in cities on mass traffic vehicles and in traffic jams.

Parameter	Value	Parameter	Value
Area size	600 m x 600 m	Bytes per data message	1400
Nodes' speed	3 m/s	Number of runs per scenario	100
Number of CenterNodes	5	Battery energy consumed in TX	25.0 mA
CenterNodes' speed	4 m/s	Battery energy consumed in RX	8.0 mA
Mobility pattern	(MRPGDC)[113]	NSM reduce factor	0.4
Chance of dynamic clustering	0.1 %	NSM reduce period	3.0 s
Bitrate	2 Mb/s	NSM minimum chance	0.02
Carrier frequency	2.4 GHz	Storage purge delay	0.5 s
Max transmission power	4 mW	Urgent category purge time	60.0 s
Signal attenuation threshold	−110 dBm	Normal category purge time	30.0 s
Path loss coefficient alpha	4	Bulk category purge time	15.0 s
MAC protocol	IEEE 802.11 WLAN	Application sending frequency	0.5 1/s
Bytes per control message	200	Simulation time	250 s

Fig. 12. Used simulation parameters

Figure 13 depicts what is the total traffic generated by the nodes. The amount of bytes is calculated as a mean of control + data traffic for a single node measured from the network layer of the nodes. A clear trend can be seen as the SAF model generates less traffic with a quite linear ratio to the number of nodes present in the network. This is mainly because SAF sends much less control messages. The epidemic model instead peaks at 40 nodes, after which the restrictions of resources starts to have an effect. Next, we inspect what are the successful delivery rates of the models. Starting with only ten nodes, both models perform equally. With 20 nodes, the percentage difference starts to grow, and from 30 nodes and beyond, epidemic routing's rate gets worse so that with 90 nodes, it is only about 2.5 per cent. The SAF model is able to keep up the delivery rate around 40 % with node counts of 20 or more.

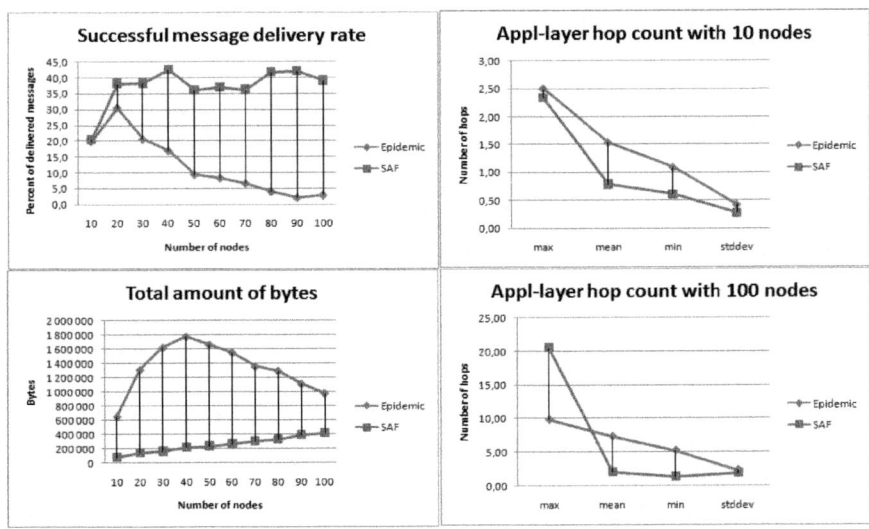

Fig. 13. Simuxlation results

In figure 13, also application (or service) layer hop counts for messages that were successfully transmitted to their destinations are depicted. The figures have been plotted with the number of nodes being 10 and 100 respectively. With both figures, SAF has smaller mean delays than with epidemic routing model. This is interesting because if the epidemic routing would work in an ideal world with endless resources, it should disseminate messages quickly as it would pass the messages blindly everywhere. However, the restrictions of storage space, messages' validity and categories have a big impact on how well that kind of approach works in a more realistic environment which has been used in the simulations. One should still notice also that in more dense scenarios, SAF can suffer from higher maximum hop counts. That is an indication that the algorithm seems to favor forwarding paths where a message has to travel through more nodes. However, from a bigger perspective does not seem to be large problems as the total network traffic numbers are smaller with SAF.

3.2 Discussion

The evaluation of the situated service oriented messaging has been carried out using OMNeT++ discrete event simulation environment. First the SAF module has been developed, and then the SOF module is investigated. The first simulation results indicate that SAF can outperform Epidemic Routing in many aspects as discussed in section 3.1. The SOF module does not require any complex, time consuming computations nor extensive memory storage capabilities, but only simple decisions and small memory buffer. Therefore, the realization of algorithm is scalable also to small mobile devices. It is estimated that the SOF further optimizes the messaging performance, and after realizing it, the final performance evaluation of the situated service oriented messaging can be performed. Especially, the case where a large number of nodes spread unevenly with pertinent nodes strewn among them randomly is interesting, because it is assumed that the messages will be more efficiently distributed to the pertinent nodes while causing less disturbance the non-interested nodes.

The evaluation of the situated service oriented approach still left open how the solution operate in more complicated situations such as different topologies, mobility, multiple radio technologies, strong security requirements etc., and especially the scalability properties of the solution.

4   Conclusions

The key contribution of the paper is the concept for situated service oriented messaging and novel algorithms for biologically inspired opportunistic networks. Simulations are applied to evaluate the usefulness of the concept. The solution utilizes different contextual information sources to create and update a view of the communicational situation. Smart diffusion of relevant control data between neighbouring nodes using novel swarm intelligence based method enables spreading of information only to the interested nodes without unnecessarily disturbing the non interested nodes. The provided methods are then applied when deciding what to do for the incoming messages in each individual user node. The local self-organization capabilities, processing and decision making enables better scalability of the messaging. The evaluations done against epidemic routing protocol indicate that the proposed solution lowers the amount of transmissions in the network, thus reducing precious resource usage in the nodes. This is achieved without introducing further delays or deteriorations in the message delivery ratio.

The evaluation of the situated service oriented approach still left open how the solution operate in more complicated situations such as different topologies, mobility, multiple radio technologies, strong security requirements etc., and especially the scalability properties of the solution.. In the next step of this research, the aim is to simulate more complicated situations of the situated service oriented messaging and especially evaluation its' scalability properties.

Acknowledgments. This work has been partially supported by the EC within the framework of the BIONETS project (see www.bionets.eu). The authors would like to thank the project partners for the good co-operation in the project. In addition, our thanks are provided to OMNeT++ community for the working simulation platform.

References

1. Fall, K.: A delay-tolerant network architecture for challenged internets. In: SIGCOMM 2003: Proceedings of the 2003 conference on Applications, technologies, architectures, and protocols for computer communications, pp. 27–34. ACM Press, New York (2003)
2. Farrell, S., Cahill, V., Geraghty, D., Humphreys, I., McDonald, P., et al.: When TCP Breaks: Delay-and Disruption-Tolerant Networking. IEEE Internet Computing 10(4), 72–78 (2006)
3. Cerf, V., et al.: Delay-Tolerant Networking Architecture. RFC 4838, IETF (2007)
4. Pelusi, L., Passarella, A., Conti, M.: Opportunistic Networking: Data Forwarding in Disconnected Mobile Ad hoc Networks. IEEE Communications Magazine, 134–141 (November 2006)
5. Ott, J., Kutscher, D., Dwertmann, C.: Integrating DTN and MANET routing. In: CHANTS 2006: Proceedings of the 2006 SIGCOMM workshop on Challenged networks, pp. 221–228. ACM Press, New York (2006)
6. Zhang, Z.: Routing in intermittently connected mobile ad hoc networks and delay tolerant networks: overview and challenges. IEEE Communications Surveys & Tutorials 8(1), 24–37 (2006)
7. Vahdat, A., Becker, D.: Epidemic routing for partially connected ad hoc networks. Research report, Duke University (2000)
8. Burns, B., Brock, O., Levine, B.N.: MV routing and capacity building in disruption tolerant networks. In: Proc. Infocom. IEEE, Los Alamitos (2005)
9. Widmer, J., Le Boudec, J.Y.: Network coding for efficient communication in extreme networks. In: Applications, Technologies, Architectures, and Protocols for Computer Communication, pp. 284–291 (2005)
10. Musolesi, M., Hailes, S., Mascolo, C.: Adaptive routing for intermittently connected mobile ad hoc networks. In: Sixth IEEE International Symposium on a World of Wireless Mobile and Multimedia Networks (WoWMoM), pp. 183–189 (2005)
11. Leguay, J., Friedman, T., Conan, V.: DTN routing in a mobility pattern space. In: Applications, Technologies, Architectures, and Protocols for Computer Communication, pp. 276–283 (2005)

12. Latvakoski, J., Hautakoski, T.: Situated Message Delivery for Opportunistic Networks. In: ICT Mobile and Wireless Communications Summit 2008, Stockholm/Sweden, June 10-12 (2008)
13. Di Caro, G., Dorigo, M.: AntNet: Distributed Stigmergetic Control for Communications Networks. Journal of Artificial Intelligence Research (JAIR) 9, 317–365 (1998)
14. Pham, D., Ghanbarzadeh, A., Koc, E., Otri, S., Rahim, S., Zaidi, M.: The bees algorithm–a novel tool for complex optimisation problems. In: Intelligent Production Machines and Systems: 2nd I* PROMS Virtual Conference, July 3-14, p. 454. Elsevier Science Ltd., Amsterdam (2006)
15. Szabó, S. (ed.): The initial mathematical models of new BIONETS network elements and algorithms. BIONETS (IST-2004-2.3.4-FP6-027748) Deliverable D1.3.1 (2006)

Simulation and Implementation of the Autonomic Service Mobility Framework

Janne Lahti[*], Helena Rivas, Jyrki Huusko, and Ville Könönen

VTT
Kaitoväylä 1, P.O.BOX 1100, FIN-90571, Oulu, Finland
{janne.lahti,helena.rivas,jyrki.huusko,ville.kononen}@vtt.fi

Abstract. The increased traffic load, proliferation of network nodes and, in particular, wireless user devices, and the boom in user services and exponential growth of information stored in content distribution networks (CDNs) have brought new challenges for current networks. One major challenge has and continues to be efficient load balancing and information access. The topics have been well studied for wired networks for, for example, process load balancing in distributed computer networks with migration. However, the wireless networks and ubiquitous computing environments create new limitations and additional requirements to perform service or process migration. In this paper, we present a simulation case and proof-of-concept implementation for service mobility as a part of the BIONETS service evolution process with the aim of optimizing service penetration in a pervasive computing environment and balancing the load in the system caused by the high service utilization rate.

Keywords: Migration, mobility, service architecture, Internet, utility function.

1 Introduction

The current network environment can already be characterized by an extremely large number of networked devices possessing computing and communication capabilities. The trend is towards a ubiquitous network environment where the networked embedded devices integrate seamlessly into everyday use. At the same time, the networks are becoming increasingly information and service centric, with the conventional communication and service provisioning approaches starting to become ineffective as they fail to address the device and service heterogeneity, the huge number of nodes with consequent scalability, node and network mobility and device/network management well. From the communication point of view, the scalability and management issues, in particular, decimate the possibility of arranging a global always-connected network infrastructure. In other words, the global network starts to resemble an archipelago of network islands. In such a pervasive, decentralized computing environment, one of the key challenges is to arrange efficient load balancing, and guarantee service penetration and user access.

---

[*] Corresponding author.

E. Altman et al. (Eds.): Bionetics 2009, LNICST 39, pp. 65–76, 2010.
© Institute for Computer Sciences, Social-Informatics and Telecommunications Engineering 2010

In order to tackle the problems of pervasive computing environments, the BIO-NETS project has proposed the architecture solution SerWorks, which incorporates service and network architectures and benefits from the biologically inspired communication paradigms [1, 2]. According to the BIONETS concept, services and service management are autonomic, and services evolve to adapt to the surrounding environment, just as living organisms evolve by natural selection.

One solution to improving system efficiency, balancing load in the system and optimizing service penetration in order to guarantee access for the majority of users is to utilize the concept of service mobility. Here, service mobility is described as the migration of the service or part of the service between nodes. The service mobility concept includes not only the migration procedure but also the selection of the service, possible replication or deprecation of the original service, migration of service implementation, its execution state and runtime data, adaptation to the target platform and handover of user associations.

Several similar solutions have been introduced on other biologically inspired communication platforms. Nakano and Suda in [3] and [4], for example, have introduced a network framework based on software agents to model the services. In addition, Suzuki and Suda have presented support for autonomic service management on [5] a middleware platform on top of a JAVA virtual machine. The main difference between BIONETS's approach and these solutions is the management of the services and implementation of the decision logics. The BIONETS platform utilizes so-called mediator entities for management and decision-making, e.g., for service migration. The mediators in BIONETS SerWorks are service external, located at each node, and one mediator can serve several individual service components where, as in the agent-based systems, each agent needs to implement algorithms for the decision logic itself. With BIONETS's "semi-centralized" system, the complexity of the services can be minimized by introducing the management functionalities outside the service and, with generic interfaces, it is possible to also provide evolution mechanisms for "legacy" services without re-implementing those as software agents.

In recent years, much work has also been carried out on load balancing using migration. In these cases, the migration is usually referred to as process migration for distributed computer systems [6] and the decision-making processes mainly consider only the CPU and memory load. Bearing in mind the pervasive computing environment and wireless ad-hoc type networks, the load balancing also needs to be taken into consideration, for example, user preferences and network conditions as well as system resources. We have defined a utility function for this to aid the migration decision, and we argue that with such an approach, the system is implementable, and by utilizing controlled service mobility, it is possible to achieve improved quality of service (QoS) also in a disconnected ad-hoc network environment.

The rest of the paper is arranged as follows. In Section 2, we present the system model for service mobility support in BIONETS. Section 3 concentrates on simulation modelling and results. Section 4 discusses proof-of-concept implementation for IP-based networks, and, finally, in Section 5 we draw the conclusions from the results and introduce our future work.

2   System Model

The ecosystem, similar to that presented in the BIONETS project [1], in which the Services live is illustrated in Figure 1. The decentralized and mobile ad-hoc network environment can be characterized by the temporally formed islands of devices that are dependent on the movement patterns of the user. On top of this networking infrastructure we are establishing a Mediator plane that provides autonomic control functionalities, network connections and other platform resources to services. Above the Mediator plane we have a user-centric service environment that allows for seamlessly integrating services of different devices in the user's surroundings to fulfil the user's needs. The atomic services (Service Cells) running on top of the platform can be any kind of small application providing certain functionality to other services or users. As the platform provides all the functionalities related to autonomic migration, the service only needs to implement the basic life-cycle controlling interfaces for starting, stopping, migrating, replicating and deprecating the service. With these interfaces the mediators can control the life cycle of the services located in the node.

The network topology is based on mobile ad-hoc connections between nodes. The devices cannot assume any backbone connection, and all the communication is done with the nodes inside the local connection range. The devices connected over local short-range links have the capability to locate, communicate and provide services for each other.

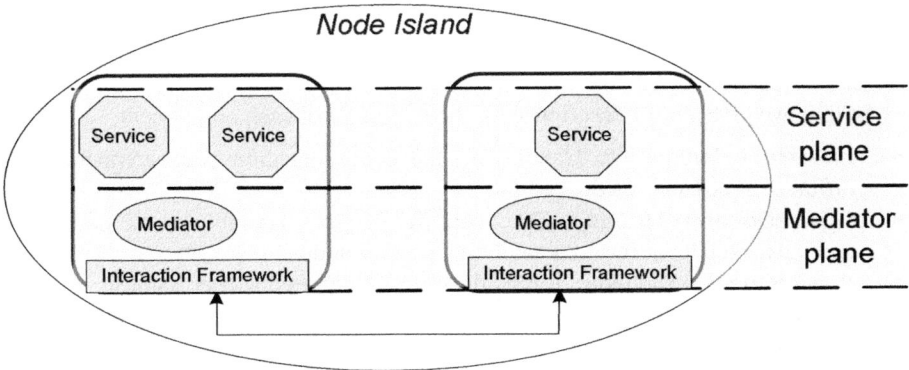

Fig. 1. Service and Mediator planes in U-nodes

The runtime architecture of our proposed service mobility framework is presented in Figure 2. The runtime platform operates on top of the nodes' (devices) operating system and provides the functionalities needed for service mobility. We have chosen a two-tier approach: we have a platform providing the autonomic functionalities for services on top of which we have the service execution layer where the services operate. The platform consists of the following functional blocks: *NodeMediator, InteractionFramework, ServiceDiscoveryMediator, ServiceCreationMediator, MigrationMediator, ServiceRequestHandlingMediator, ServiceMediator, ServiceExecutionRepository, ServiceBuffer and Node Information Repository.*

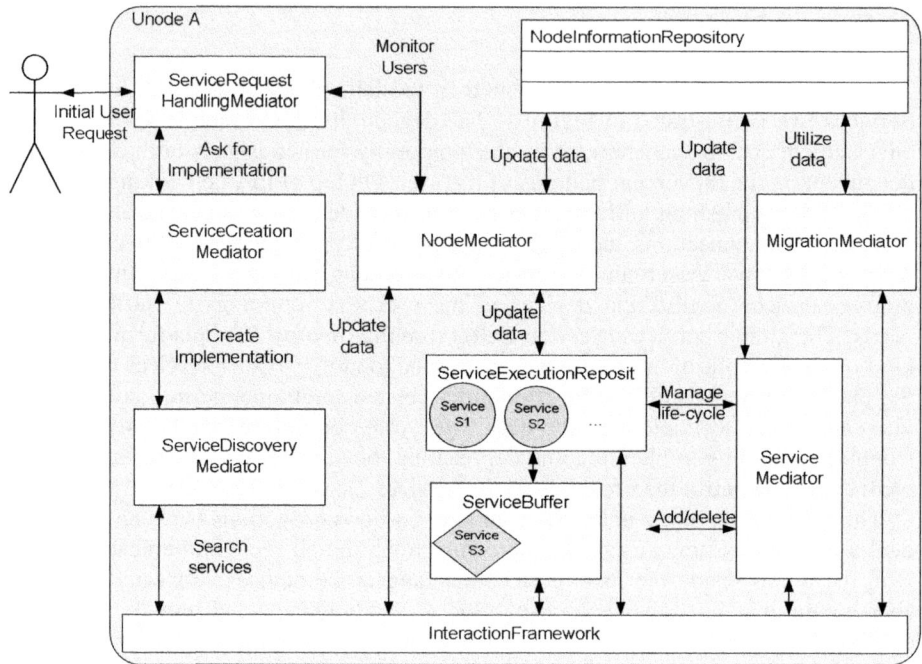

Fig. 2. The main components of BIONETS SerWorks Architecture needed to support service migration

Below, we give a brief description of the main role of each component.

- *InteractionFramework:* Handles all the communication between mediators in different nodes and inside the node.
- *ServiceDiscoveryMediator:* Discovers the services available in the node island, and makes the local services available to other nodes.
- *ServiceCreationMediator:* Builds the core for processing user requests.
- *ServiceRequestHandlingMediator:* Builds the interface to the user.
- *ServiceMediator:* Suspends service execution, migrates runtime service data and state to a destination node, initiates new service with restored state on the new destination and resumes application execution.
- *MigrationMediator:* Enables the movement of service application logic between different nodes.
- *NodeMediator:* Gathers and distributes the necessary information for the decision logic needed in the migration process, e.g., node fitness, CPU, battery, memory, speed or signal strength, etc.
- *NodeInformationRepository:* Keeps the information gathered by the Node Mediator.
- *ServiceBuffer:* Stores the services that are not currently executed in the node.
- *ServiceExecutionRepository:* Service execution environment. When an executing service is suspended, it is moved to the ServiceBuffer.

3 Simulation Model and Results

3.1 Simulation Model

The purpose of the simulation work is to model the service mobility in a BIONETS environment in which devices (nodes) move in a limited geographical area and dynamically form Service Islands with nearby nodes. Some of these nodes contain services offering limited sets of services (Service Cell A, B, C and D). In the model there are also entities (users or other services) connected to the nodes using one or all of those services.

The purpose of the simulations is to obtain knowledge about how the selected technologies and solutions would work in real implementations. We compare different decision-making mechanisms, migration models and overall service architecture variations to find out which ones are best suited for BIONETS-like environments. The purpose is to find solutions that ensure the best "Service penetration" for popular services with the minimum overhead cost. For example, we will look into whether the decision logic is more beneficial for implementation in the Service Cell or in the Mediator. The networking aspects related to the Service Migration are left outside the scope of the simulation. The simulation model does not implement the network level functionalities. The connections between different nodes are handled by the InteractionFramework (IF) component. The IFs of different nodes can "see" each other and form logical connections only when the nodes are in the same Node Island (NI), thus inside the given connection range.

The simulation model is implemented on top of the MASON simulation toolkit [7]. MASON is a fast discrete-event, multi-agent simulation library core in Java designed as the basis for large custom-purpose Java simulations. On top of MASON, we implemented the core BIONETS components presented in Section 2 (see Figure 2) that are required for service mobility purposes. The Mason toolkit executes all the components (U-nodes and Service Cells) at every step, with one step representing one second in real time.

In the simulation model, U-nodes move randomly in an X m*X m playground (City or Open field). The movement of nodes is based on the random waypoint mobility model [8] with an enhancement of "service gravity". This movement model provides semi-random movement with nodes picking a target point somewhere inside the circle with radius r and starting to move towards that point at every step. In some cases, we enhance the movement model with a "gravity factor", which makes the nodes less likely to leave an area that is providing services used by that node.

The U-nodes contain Service Cell components' and/or ServiceUser components. The ServiceUser mimics the real-world service users (user or other service) and is fixed to their host nodes (in certain cases the user may "change the device", i.e., move to another node). The services can (depending on the scenario) migrate freely between U-nodes. The U-nodes communicate with each other through the InteractionFramework (best effort messaging). The network part of BIONETS is not implemented in the simulation model. The U-nodes have a connection range and form NodeIslands with other nodes inside the range, enabling the InteractionFramework to exchange messages.

For reasons of simplicity, we do not have separate variables for each U-node's resource (e.g., CPU, memory and disk space), but model all these basic resources as general variables: Runtime Resource (*rr*) and Static Resource (*sr*). Each U-node has an amount of *x* resources *rr* and *sr*. The rr models the runtime resources consumed by services such as CPU and Memory load, and the sr models the static resource consumption such as disk space. Each hosted Service Cell consumes a certain amount of static resources from the host U-node when installed and frees it when it is uninstalled. In contrast, the runtime resources are consumed by Service Cells when executing a service response, and these are reset at each step. Similarly, the Service Cells have the variables Static Resource consumption (src) and Runtime Resource Consumption (rrc), modelling the amount of static resources required by the Service Cell when installed to a U-node and the amount of runtime resources consumed by the Service Cell per service response.

A more detailed view of simulation model implementation is presented in Appendix 1, which presents a sequence diagram of service migration in the Simulation Model.

3.2 Simulation Results

The primary purpose of a simulator model was to provide a (somewhat simplified) view of a model to be tested and studied. In the first phase of the research work we created a simple hypothesis: "It is possible to achieve improved service penetration in a disconnected ad-hoc network environment utilizing controlled service mobility." In order to evaluate the hypothesis, we envisaged a three-parted simulation study to set the base line for future more advances simulation studies. The simulation contained three different scenarios presenting different migration models: *Static services*, *Viral distribution* and *Controlled mobility*.

In the Static services scenario, the services laid completely static in their host nodes. The purpose of the scenario was to provide a reference point where, much like with the current systems, the services do not migrate and thus provide weak service penetration, but at same time require minimum resource consumption. In the Viral distribution scenario, when the U-node hosting service x comes within the connection range of another U-node, the node migrates the service to the new U-node. After the U-node is "infected" with the service x it also starts to distribute it to other U-nodes inside the connection range. The purpose of the scenario was to provide another opposite reference point where the services spread uncontrollably to new nodes resulting in very good service penetration, but with the cost of inflated resource consumption.

In the Controlled mobility scenario, the services migrate, applying the rules given by the MigrationMediator's decision process. The assumption was that the service penetration is better in this scenario than in the Static scenario, but with reasonable resource consumption. In order to evaluate the different scenarios, we run extensive simulations using the presented simulation model (Section 3.1). We set the square size L to 2 km, the communication range R to 50 m and the speed of the U-node V to 4 m/s. In each simulation run, we injected one hundred U-nodes into the environment in random locations. We also implemented four different Service Cell types (SC1, SC2, SC3, SC4), each providing simple calculation tasks (SC1 = add{x, y}, SC2 = subtract{x, y}, SC3 = multiply{x, y}, SC4 = divide{x, y}), and injected 10 of each into the random U-nodes. We also randomly injected 20 Users into the nodes. The Users were set to create queries to one (random) service type at random every 5 steps (seconds). We measured the number of successful replies (SRN) to the User on average for the sent service request in each simulation run, thus giving a reasonable presentation of the service penetration among U-nodes. We also measured the average resource consumptions for static resources (ASR) and runtime resources (ARR). We scaled all the numerical results to [0.1] for better comparisons.

We ran the simulation 20 times with each simulation setting. Each simulation run lasted 10000 steps (seconds). From histograms presented in Figure 3 we can see how the number of successful replies varied in different scenarios. Figures 3a, 3b and 3c present the distribution of measured SRN values for separate simulation runs in each case. As expected, the best service penetration was in the Viral distribution scenario in which the Users received responses to sent service queries on average about 49.6% of the time. In the Static scenario, the percentage was significantly lower at 10.4%. In the Controlled migration scenario, in which the services migrate according to predefined rules, taking account of the resource load, the percentage was 29.8%.

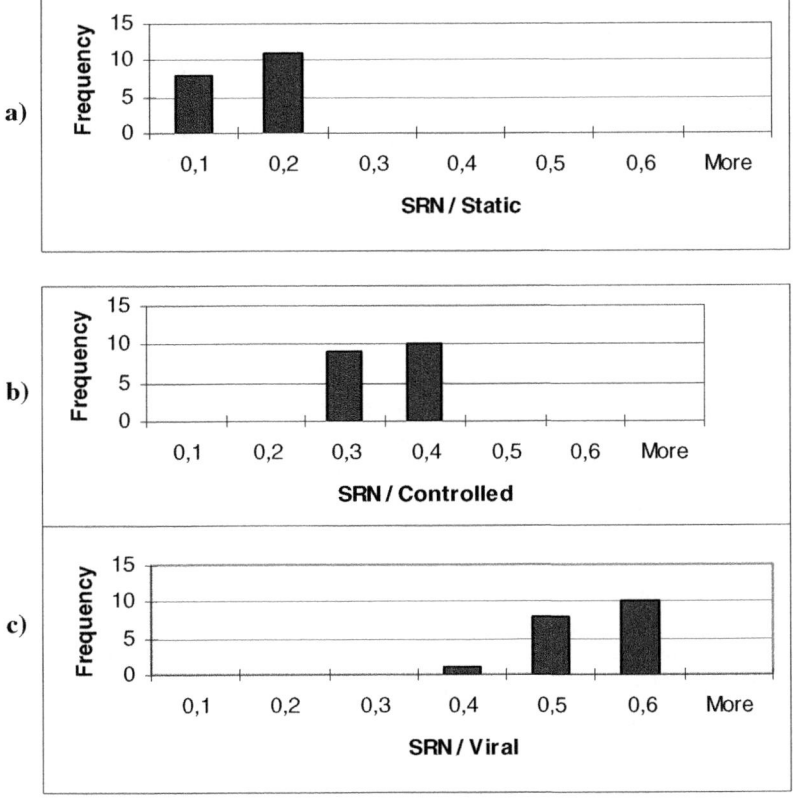

Fig. 3. Distribution of SRN values for each case

The average resource consumptions for same simulation runs are illustrated in Figure 4 below. As we can see, the resource consumption is somewhat reversed compared to the SRN values. The AVG variable presents the average of both ASR and ARR values. We can compare the "goodness" of each scenario by calculating a value $Q = SNR / AVG$ for each scenario (bigger values are better). For the Static services scenario, we get $Q = 0.846$. For the Viral distribution scenario, we get $Q = 0.728$. Finally, for Controlled mobility, we get $Q = 0.937$, which supports the original hypothesis that we can achieve better service penetration with reasonable resource consumption with controlled service mobility.

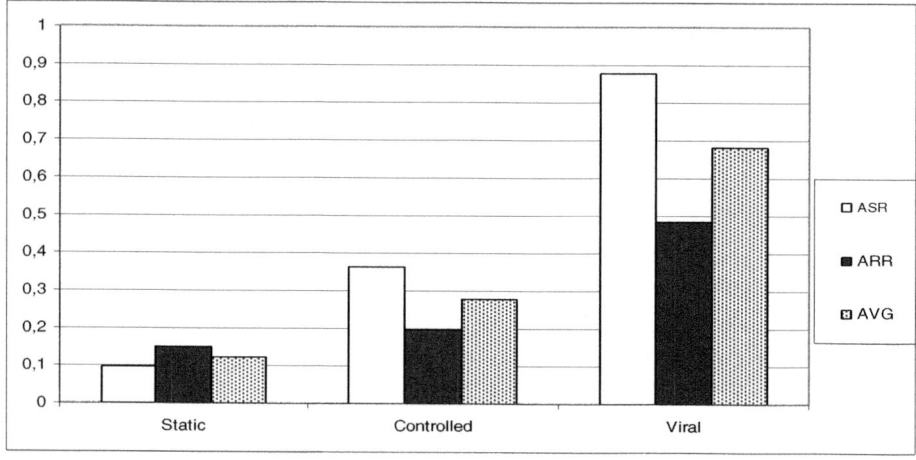

Fig. 4. Resource consumption values for different cases.

4 Prototype

The prototype model implements the basic BIONETS components required for service mobility purposes. The prototype implementation was developed on top of the J2EE platform and the services were implemented as web services. PC/Linux was selected as the implementation platform mainly because of its flexibility in collecting network information. It also provides easier access to system parameters required for decision-making. SIGAR API [9] was used to access the system information required for decision-making. This SIGAR (System Information Gatherer and Reporter) is a cross-platform, cross-language library and command-line tool for accessing operating system and hardware-level information in Java, Perl and .NET technologies. The SIGAR API enables the developed platform to also run on top of Windows OS. Glassfish [10] was chosen as the web server due to its programmatic Application Server Management Extensions (AMX) [11] API. AMX is a superset of the JSR 77 interfaces built on JMX, which simplifies and smoothes out the management and monitoring process. The information gathered with SIGAR (system information, e.g., memory usage, CPU consumption, networking, etc.) and AMX (web-services information: response times, throughput, total number of requests, faults, etc.) is collected in the Node Information Repository.

In the following, we present simplified decision logic that we implemented for real devices based on the research work and simulation results. We want to emphasize that the parameterization and utility functions here are only preliminary, though they are general enough to cover many real world tasks and will be extended later.

Each time a node makes its decision, it has a fixed set of options. The fitness utility value is computed for each possible option using the utility functions (see below). An option with the highest utility value, that is the best fit for this purpose, is then selected for execution. In a similar way to the simulation model, we implemented four simple integer calculators (add, subtract, multiply and divide) as services, which migrate over the developed platform running inside real devices (Linux OS laptops).

Below is an example of a simple utility function for calculating the fitness of Service Cells based on the memory usage and operation-related CPU. A detailed explanation of the migration process implemented in the prototype is presented in Appendix 2.

Parameters:

> Number of requests (SC_{NoR}), size of service: memory allocation
>
> required (SC_{MA}), operation-related CPU (SC_{CPU}) and memory usage
>
> (SC_{MU}), user evaluation (SC_{UE}), battery (U_B) and free memory (U_{FM}).

Utility function for SC:

> c(SC_{NoR}, SC_{MA}, SC_{CPU}, SC_{MU}, SC_{UE}, U_B, U_{CPU}, U_{FM}), $c \in L$
>
> c is a function of the parameters of a service cell and the platform on which
>
> the service is running. L is a partially ordered set, e.g., the unit interval [0,1].

Example utility function:

$$c(SC_{CPU}, SC_{MU}) = \alpha(U_{CPU} - SC_{CPU}) + (1 - \alpha)(U_{FM} - SC_{MU})$$

> α is a parameter in the unit interval [0,1], which weighs the importance of memory and CPU usage.

5   Conclusion and Future Work

In this paper, we presented the service mobility framework in the BIONETS concept together with the results of the service level simulations with a MASON simulator and proof-of-concept demonstration for an IP network with PC hardware. The framework can be utilized for balancing the service load and resources, and to optimize the service penetration in and between the network islands, when the connection is not always guaranteed. We discussed basic principles of the service ecosystem and presented a system model for service mobility support in BIONETS. We also presented a simulation model for service migration and a simplified prototype implementation. Finally, we presented preliminary simulation results. The main results confirmed that the controlled service mobility can provide better service penetration with reasonable resource and energy consumption.

The next step in our research is to better optimize the service mobility management by applying more extensively, for example, game theory and learning capabilities to the decision-making. One possible approach is also to utilize more efficient mobility triggering and define the triggering events needed for the mobility management as presented in, for example, [12] for node mobility. In addition, we are aiming to continue the prototyping activities by applying the service mobility framework to, for example, energy-aware cloud computing and distributed systems, implementing the more complex utility functions for decision logics in order to also better the network conditions and provide more extensive results from the hardware implementation testing.

Acknowledgments

The authors would like to thank all their colleagues in the EU IST FP6 BIONETS project. The research was partially funded by the European Commission.

References

1. Chlamtac, I., Miorandi, D., Steglich, S., Radusch, I., Linner, D., Huusko, J., Lahti, J.: BIONETS: Bio-Inspired Principles for Service Provisioning in Pervasive Computing Environments. In: Di Nitto, E., Sassen, A.M., Traverso, P., Zwegers, A. (eds.) At your service: service engineering in the Information Society Technologies Program. MIT Press, Cambridge (2008)
2. Miorandi, D., Huusko, J., De Pellegrini, F., Pfeffer, H., Linner, D., Moiso, C., Schreckling, D.: D1.1.3/3.1.3 Serworks architecture v1.0. BIONETS Deliverable, D1.1.3/3.1.3 (2008)
3. Nakano, T., Suda, T.: Adaptive and Evolvable Network Services. In: Deb, K., et al. (eds.) GECCO 2004. LNCS, vol. 3102, pp. 151–162. Springer, Heidelberg (2004)
4. Nakano, T., Suda, T.: Self-organizing network services with evolutionary adaptation. IEEE Trans. on Neural Networks (2005)
5. Suzuki, J., Suda, T.: A middleware platform for a biologically inspired network architecture supporting autonomous and adaptive applications. IEEE Journal on Selected Areas in Communications 23(2), 249–260 (2005)
6. Milojičić, D.S., Douglis, F., Paindaveine, Y., Wheeler, R., Zhou, S.: Process migration. ACM Computing Surveys (CSUR) 32(3), 241–299 (2000)
7. Luke, S., Cioffi-Revilla, C., Panait, L., Sullivan, K.: MASON: A New Multi-Agent Simulation Toolkit. In: Proceedings of the 2004 SwarmFest Workshop (2004)
8. Yoon, J., Liu, M., Noble, B.: Sound mobility models. In: Proc. of ACM MobiCom, San Diego, CA (2003)
9. SIGAR API, http://www.hyperic.com/products/sigar.html
10. GlassFish, https://glassfish.dev.java.net
11. Appserver Management Extensions, https://glassfish.dev.java.net/javaee5/amx
12. Mäkelä, J., Pentikousis, K., Majanen, M., Huusko, J.: Trigger management and mobile node cooperation. In: Katz, M., Fitzek, F.H.P. (eds.) Cognitive Wireless Networks: Concepts, Methodologies and Visions – Inspiring the Age of Enlightenment of Wireless Communications, pp. 199–211. Springer, Heidelberg (2007)

Appendix 1: UML Sequence Diagram of Service Migration Implementation in the Simulation Model

Bellow we present the UML diagram showing the migration process implemented in the simulation implementation. In this figure, we show the process sequence for a scenario where the NodeMediator of the U-Node triggers a service migration of Service Cell X to a new host. The MigrationMediator offers the X to U-Node B (which accepts it) and then migrates the Service Cell to a U-node B. It also shows the interaction between *Mediators* and the exchange of messages between them.

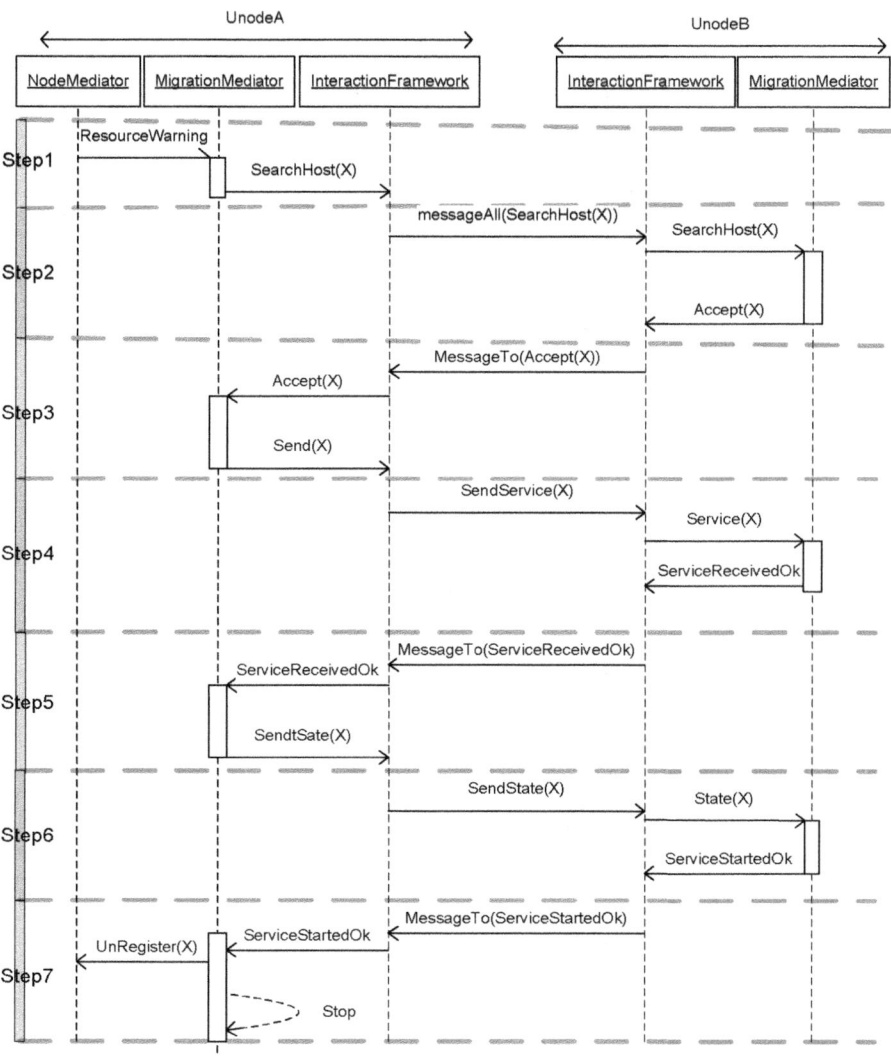

Appendix 2: UML Diagram for a Prototype Implementation

Bellow we present the UML diagram showing the migration process implemented in the prototype implementation. In this figure, we show the *requestMigration* of U-Node A to U-Node B to migrate its *ServiceX* to U-node A. It also shows the interaction between *Mediators* and the exchange of messages between them.

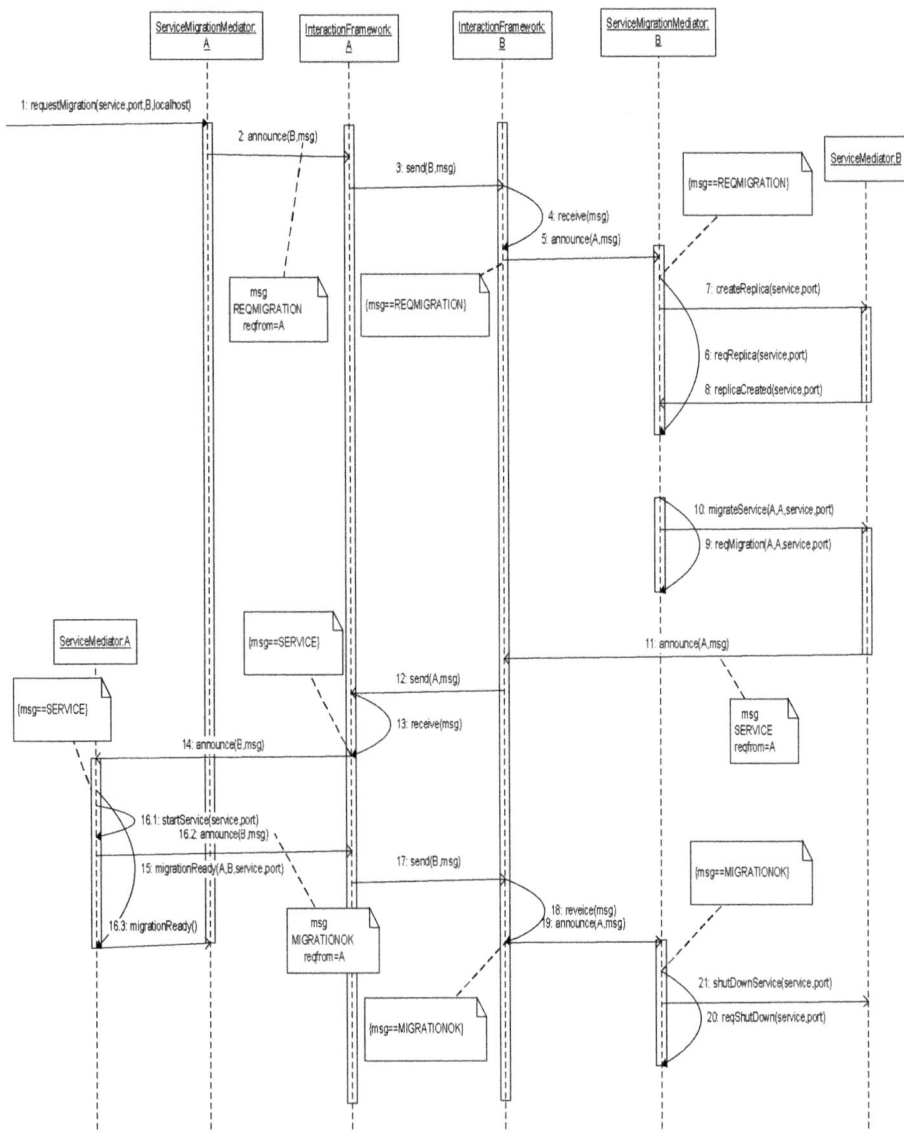

BIONETS Economics and Business Simulation: An Alternative Approach to Quantifying the Added Value for Distributed Mobile Communications and Exchanges

Silvia Elaluf-Calderwood and Paolo Dini

Department of Media and Communications
The London School of Economics and Political Science
Houghton Street, London WC2A 2AE
{s.m.elaluf-calderwood,p.dini}@lse.ac.uk

Abstract. This article presents current research work on the development of BEBS (BIONETS Economic and Business Simulation Model). This model is used to illustrate how in pervasive agent-based networks dynamic agents engage in distributed mobile communication exchanges that carry a potential, non-quantified value to be determined. This value can be monetised either as advertisement or as share of revenue profit for an external content or service provider wishing to distribute information over the network. The strength of the simulation model is to enable different ways in which the value of communication events can be quantified and that are not normally considered in conventional business approaches. Our simulation can therefore facilitate the exploration and development of alternative business models based on heterodox economic perspectives such as the economics of sharing, gift economy, and economic sociology applied to mobile networks.

Keywords: social networking economics, alternative business models, token exchange, economics of sharing, mobile networks.

1 Introduction

"There is no recipe for the successful use of social tools. Instead, every working system is a mix of social and technological factors" [21]

The use of mobile devices and the 'Web 2.0' phenomenon have opened virtual and real windows for ubiquitous and seamless social communications [1], which are paradoxically both continuous and fragmented [25]. Although many research efforts have focused on understanding the social aspects of networking using mobile devices [3], the understanding of how these networks can be valued in terms of money is an unfinished research topic. Social networking draws its foundation from Granovetter's theory built around the concept of "the strength of weak ties" [11]. The brilliance of this theory helps understand how some individuals can be nodes of convergence for many other individuals that otherwise would have nothing in common [2].

E. Altman et al. (Eds.): Bionetics 2009, LNICST 39, pp. 77–87, 2010.

In many ways the use of mobile devices has taken the lead role in introducing new social organizational forms [19]. The combination of mobile devices with networking tools such as Facebook, Twitter, Hi5, and other similar web-based applications – collectively referred to as Web 2.0 – has created a chorus of approval from social experts [18, 24] for the apparent success of social networking as a means to achieve a ubiquitous connected society. Analogies can be made on mobile phone use spread to the study the fundamental spreading patterns characterizing a mobile virus outbreak [14]. Social networking in fact shifts the generation of media from the technology to the user and the content users generate [2, 12].

This shift has been accelerated by the rapid evolution of mobile devices that are adopted by users, because these devices have features that can perform and enhance seamless communications. Users are eager to transfer and exchange many forms of data on the virtual networks accessed through mobile devices [20]. Some of these communication exchanges open up new social networks; others reinforce already existing ones [26].

Although this exchange of data (e.g. user profiles, video files, music files, documents, games, etc) benefits social networking, in economic terms the increase in volume of data transferred has a cost that operators quantify. The emergence of pervasive and interconnected computing devices that give rise to a dynamic network topology in which distributed content can become focalized and spontaneously shared brings new opportunities for economic exchange. The patterns of communication are based on social networking and bio-inspired models such as epidemic spreading and gossiping metaphors [10, 13], where trust is all-important and reputation keeps local trust values above other values in the network [4]. Thus, social networking and mobile applications have created a new real and virtual space in which the traditional models of revenue might need to be re-thought [16, 17].

Telecom operators are the main referees when applying a distribution model to determine these costs. However in pervasive and distributed agent networks a significant volume of data transactions is distributed by and between nodes and not necessarily linked primarily to the telecom operator's backbone. As users turn to providing more and more content, agent networks become an interesting case of study from an economic point of view. The value or cost for these distributed communication exchanges are not well quantified under current business models.

Current business models base their revenue calculations on conventional advertisement. The fact that this is an emergent technology is responsible for the dearth of economic studies looking into this subject. There is a wide berth between theory and applications. Current attempts to estimate these values have been undertaken by e.g. charities trying to estimate the value of using social networking for their campaigns [9]; others rely on models in which each user assigns the relative value to the communications within the network, as Twitter users discuss [17, 20, 23].

In this paper, the BIONETS network provides the context in which potential value of exchange are discussed. For this purpose Benkler's [5, 6] proposal about value-added distribution and the emergence of sharing as a mode of production is used as the basis for the simulation developed, while focusing on the economics of sharing for the distribution of content. The idea behind the simulation is to provide a benchmarking tool that can support the development of distributed applications over a disconnected network of mobile devices and sensors within the context of the BIONETS EU project. In particular, we aim to probe our assumptions about where the value of such communications is, through different scenarios and models for its quantification.

2   Bionets Communication Exchanges

A network such as BIONETS, where the emphasis on communication exchanges and engagements is influenced by an evolutionary bio-inspired framework, that is node-based and distributed, is an ideal environment where to try to determine the values of communication exchanges and how to quantify them. In many ways the BIONETS case fits the requirements as many of its processes are seen as an industrial mutation that incessantly revolutionizes the economic structure from within, incessantly destroying the old one, incessantly creating a new one [8, 20].

Figure 1 shows the actors found in the BIONETS architecture. The actors aim to facilitate communication exchanges of diverse types, and consider for those five main components:

1. Technology expressed by the boundaries determined by device manufacturers and network equipment vendors.
2. Services: the discussion in this document focuses on value-added services, content and applications that users or other network devices can access through the BIO-NETS mobile network: the symbiotic relationship between content providers (individual or networked), application providers, and payment agents/or exchange agents.
3. Network, U-nodes, T-nodes that effectively work at the nuclear level as mobile network operator or ISP.
4. Regulation, protecting the privacy of users, regulating the market and legislation and other requirements for service provision
5. Users demand determines the success or failure or evolution of BIONETS services or applications.

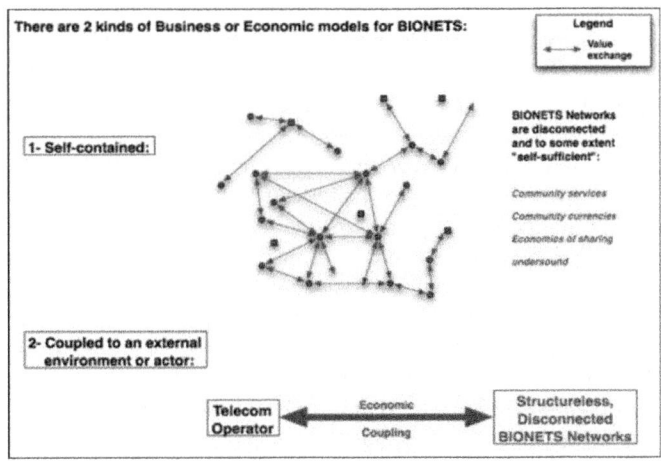

Fig. 1. BIONETS Networks

3 Distributed Mobile Networks: Economic Paradigms

In the context of BIONETS there are a number of economic paradigms that need further explanation in order to understand the working model for BEBS. At the core of the work on usage scenarios presented by several BIONETS partners there are two main networks of interaction in which business models might develop. The first network is a self-contained, 'disconnected' and distributed BIONETS network of nodes (actors) able to exchange data with each other using the token values of alternative trading systems – where no actual money changes hands – whilst the other network is that used by the telecom operators, in which economic revenues are based on more conventional business models. Both models are depicted in Figure 1.

Neither of these networks has an exchange rate to convert the services or transmissions within the distributed BIONETS network into revenue, nor within the conventional telecom network. The decision to build a computing model using a modelling tool to illustrate alternative business models is a way of evaluating how these exchange rates could be calculated.

The economic principles of the model are based on the use cases developed by BIONETS researchers, which can be grouped at an abstract modelling level as the ability of each node or actor to transmit (send or receive) up to five types of data (this can be changed in the model), assigning to each of these transmissions a token value for the desirability of this transaction. The data transmitted can be for example either environmental variables such as the ones presented in the Digital City scenario [7], with services that for example allow real-time access to Digital Maps or music exchanges, or a combination of all of these.

A generic model in this case tries to express the added value of the exchanges that occur in the node-based networks that are traded, for either other tokens or real monetary value, by the actors or users of the network.

For example, consider the situation in a metro station in which any agent in a certain area within the station can run BIONETS applications. Each agent can receive or send data based on their needs. Assume that at some point in time there is a number of people waiting on the platforms for trains to arrive; whilst waiting, some if not all of those people might use their mobile devices locally. Each of those devices is an agent in the business model being simulated. As agents discover other agents they might start exchanges of data that result in economic transactions based on agreed token values or evolving token values. A historical record of such transactions could be stored in an individual virtual account for each agent, thereby making possible a valuation of the desire or ability to exchange data successfully.

One type of exchange could be the case of an agent broadcasting to other agents within range a message or data file containing some kind of advertisement; all the other users within range might choose to accept or reject the sender's file; if, however, an agent decides to accept the sender's file, the sender is credited with a token value paid into their virtual account. This could be the case for a localized advertisement used to reach a small network of users: the eagerness of users or agents to accept the advert could ultimately be converted into real monetary units for the sender by, for example, the telecom operator providing the permanent network.

Over time the number of agents in the metro station changes; at some times there will be peaks of data exchange and/or number of agents, and at other times minimum or no exchange of data or no agents at all. There is a dynamic cycle, based on agents entering and leaving the metro station, wishing to exchange or trade information. Some of the evolving, bio-inspired applications from BIONETS will merge then with social networking behaviour to express over time the changing nature of these transactions.

Conventional business models cannot effectively allocate value to this type of agent network in which the exchange of data is or can be considered separate from the backbone telecom network, as is the case for example with Bluetooth. One of the aims of the modelling is to illustrate how alternative economic models can actually build up enough subjective or token-based value to make it worthwhile to develop an exchange rate for its conversion into real money.

This can be done by evaluating the total of the exchanges in the metro station, in terms of both the number of tokens and the volume of data transferred, against the telecom's valuation of volume data transfer per minute. This will allow for example potential marketing companies and telecom providers to estimate, based on the potential number of users or actors accepting a broadcast message in the metro station, the cost of advertising localised and perhaps focus-orientated advertisements.

In this way an exchange rate of sorts is established between the advertisers and telecom network providers, based on allowing marketing companies access to these networks, and a pay-off for the telecom providers who always maintain an external connection to the metro stations.

There are many cases that could be illustrated using this type of modelling, and the complexity of each model will depend on many factors; since BEBS aims to illustrate the potential of alternative economic models running on top of the BIONETS infrastructure, heuristic choices have been applied to the model, as explained in the section below.

4 BEBS Model Fundamentals

The model fundamentals are based on representing the type of exchanges illustrated in Figure 1. The modelling assumptions are:

1. The simulation focuses on the assessment and evaluation of the self-contained economic model proposed in the figure, by assigning to each data exchange or storage a token value to be summed over a certain time period for both the overall network and individual nodes.
2. Each node will have the same set of attributes. The number of attributes has been limited for the simulation to five. See list in point 5 below.
3. Each node will be both a supplier and a consumer of communication requests.
4. A node can have a limited number of connections to other nodes based on its transmission capacity.
5. Attribute list (the token value of storage is for all the files hosted at any time in a node; it is a unique value representing the operational cost of storing data):

   Node stores transmission data, token value = 1, protocol (none)
   Node distributes transmission data, token value = 2, protocol UDP
   Node stores non-transmission data, token value = 3, protocol (none)
   Node distributes non-transmission data, token value = 4, protocol TCP
   Node can send and receive streaming services, token value = 5, protocol VoIP

6. The attributes for each node define the local environmental conditions
7. A connection between two nodes is an active link for the transmission of data or music
8. At any time the communication between two nodes will have a maximum of two channels in the same direction (one for music and one for data)
9. Over time traffic on links and channels will change randomly, keeping condition 4 as their only constraint
10. Each link will have a cost/value (token value) and the simulation will sum those values over a period of time to estimate the global economic benefit generated by the economic model

4.1 The Model

The model was developed in Repast[8] using a template, adapting the code to the requirements of the scenario to be simulated. The results are calculated and displayed at each step (discrete interval) of the simulation as a dynamic graph of the aggregate value of all the nodes' transactions plotted as a function of time. The model can be run at different time interval settings. The output can be examined in graphical and numerical form, allowing comparisons when necessary.

4.1.1 Basic Concepts

The model consists of three basic concepts: agent, link and space.

1. The agent represents a person with a mobile device that makes the decision to receive or send data using one of the means specified above. Each agent behaves independently, and the model only acts as a holder for all the agents.
2. The link is the actual communication. For the purposes of this model, it has only a value from one to five, as explained above, and a type: broadcast, i.e. one agent sends to everyone within range, and each recipient then decides whether to accept or reject the transmission; or point-to-point, where the recipient of the data is specified by the sender.
3. The space represents the metro platform that is the scene for the communication. This is a 40 x 40 grid; in which filled cells represent agents. To make the simulation more realistic, agents can only communicate within a certain range (the range for Bluetooth for example is typically 10m). Movement of agents within the space was considered but has not yet been implemented. This is an acceptable approximation for relatively small data sets that are exchanged quickly relative to the rate of change of the network topology.

At each step, some agents are created to simulate their arrival at a metro platform, whilst others are destroyed, i.e. they leave or their devices are no longer transmitting. The number arriving and the number leaving are randomly distributed around the same mean. This implies that, over a long period of time, the number of agents will average the initial number – currently set to 80. However, very large fluctuations are possible, particularly as the creation of new agents does not happen at every step, which resembles the actual pattern of people arriving at station platforms. The simulation can at times approach capacity (1600); at other times it can be almost empty. The range of agent lifespans, and frequency of creation of new agents, can be configured.

If new agents are created at every step, the number of agents will tend to be even over time. If however there are a number of steps between the creation of agents, there will be greater fluctuations.

A link has a lifespan and a random value within a configurable range. The value is added to the agent only when the link dies naturally. If an agent is destroyed, any links that have not reached their natural lifespan will be destroyed and their value will not be realised. This represents the case where someone is transmitting some data, but leaves (e.g. gets on a train) before finishing the transmission, and the partial transmission is then useless.

In the case of point-to-point transmissions, the value is added to the sender and to the receiver, whilst for broadcast transmissions it is the receiver who gets the value. This is because point-to-point transmissions are typically part of two-way communication, which have value for sender and receiver, whilst in the case of broadcast data the recipients do not respond.

4.1.2 Visualising the Model
BEBS has used the Repast built-in user interface facility for graphical emulation in 2D – topological format – of the agent network and transaction model. A graph and table of the total value of the system are also shown (see Figure 2). The agents are shown as squares and the links represented by lines between them. The colour of the line represents the value and type.

4.1.3 Parameter Inputs
The aim of the model is to see how the total value of the system changes over time. It is run with various sets of parameters presented here in a vector form (parameter, name, usage in model):

4.1.4 Simulation Outputs
BEBS provides the following simulation outputs:

1. A graphical display of the nodes and their generation
2. A graphical display of the nodes' attributes and their properties
3. A display of the simulation running over time, showing links and active channels of transmission
4. The value of a transaction over the simulation period and a selected time, calculated for each actor
5. Overall value of the network over a certain period of time

The model generates an initial number of agents, whose lifespan is allocated at birth as a random number of steps between the minimum and maximum values set. Agents that die are not immediately replaced. Instead, new agents are generated at a random step interval. The number created is approximately the sum of dead agents since the last generation, but varies between 0 and the double of this number, such that the average number replaced is equal to the average number leaving.

Say, for example, agents are replaced at step 9, 5 agents die at step 10, 6 at step 11 and 10 and step 12, then are replaced again at step 13. That means that 21 die in this time. The number of agents replaced is then a random number between 0 and 42. The reasoning behind this is that, in a network, the people represented by agents tend to arrive in groups but leave individually; the distribution however can be distributed mirroring an epidemiological model [14].

With every step, each agent makes a link with a random value from 1-5 to another agent. If the agent is within a given distance, the link is accepted; if not, the link is not made. It is possible to receive any number of incoming links. Furthermore, one agent broadcasts a link with a value of 1 to all the other agents in its vicinity, which they may accept or reject.

Figure 2 illustrates a stage in the simulation run in BEBS. The green dots are the nodes or agents exchanging information. The links have different colours depending on the type of communication exchange. The broadcasting of some type of communication is shown by the links in blue: one agent sends to many a message, the number of agents accepting the message increases the value of the node sender. In the following section a summary of the main ways this programme can be used are presented.

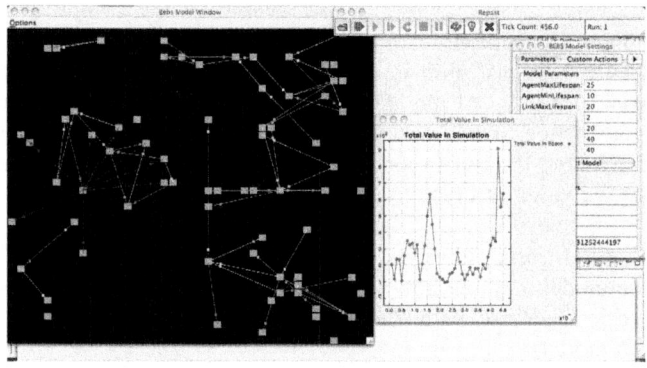

Fig. 2. Communication Exchanges in BEBS

4.2   Results

The simulation illustrates how transactions change over time and shows changes in the number of agents located in a certain area exchanging data. While currently it is not possible to attach real values to these exchanges, the flexibility of the program allows it to be executed many times, based on values that can be calibrated according to expectations determined by the agents. This will only be possible when there are running applications in BIONETS that are able to collect data about the eagerness and volume of transmissions within nodes.

Some interesting findings were obtained when running the simulation on trials using values we selected. Figure 2 shows how the program works and comments about results.

In Figure 2 the aim of the simulation is to measure the change in total value over time, and to see patterns in the variations where different parameters are employed. The total value on the y-axis is plotted against time on the x-axis . As time varies, the scale of both axes increases whilst the intervals reduce, so that the graph remains the same size. In order to compare graph shapes, a fixed number of steps can be run.

Also in the figure the value of the nodes on a certain time are saved on a .csv file that can be automatically imported by Microsoft Excel or any spreadsheet package. The most interesting configurable parameters are the maximum and minimum lifespan of a link and agent. Lifespans can be fixed within a wide or narrow range. Considering first a narrow range for both link and agent, link lifespans that are short relative to the agent lifespan tend to give a higher total value, since a larger number of short-lived links can be formed within the lifespan of the agent. Long agent and short link lifespans tend to lead to very gradual changes in value, since only a small proportion of the agents expire at any given moment. Considering now the range, a wide range of link lifespans with a narrow range of agent lifespans tends to lead to small and regular fluctuations in total value, but around a fairly steady and predictable mean. A wide range of agent lifespans, however, has the effect of radically destabilising the model. There are likely to be long periods where the total value is static, but with sudden, irregular peaks which can be very high relative to the average.

5 Remarks on BEBS

In principle this is a first attempt to develop a quantification of business models for the BIONETS infrastructure. The aim is to be able to provide, in the longer term, a model that can be interfaced with the computational output from the infrastructure development, feeding in this way real data collected in transactions that will occur when there are applications available to be deployed using the outputs from the BIONETS computational and scientific research. This is a novel approach that could be developed further and merged or integrated with the work completed by other partners in the BIONETS project. In doing so, this simulation can be a powerful tool to illustrate the economic benefits for actors derived from the exchange type of business model in any future BIONETS-enabled environment.

Until now the value of these trading networks has not been tested in mobile environments, and depending on their growth and sustainability some of these networks may in future evolve to have significant value, which will make them attractive to conventional sources of funding. E.g.: record companies releasing songs over metro stations, broadcasting companies (TV, film) distributing total or partial media files as teasers for users to develop interest in the content, users who can dedicate their resources to collect environmental variables that can be used later to tailor services or activities in a location according to user demand.

There are not many research computer models especially created to recreate social networks' economic behaviour, such as the ability to exchange information over distributed networks and allocate token values to the exchange transactions. The simulation is aimed for a business model that is not centralized; hence this piece of work contributes to enhance the understanding of these types of networks, examples of this type of work can be found in the work of Harwood [12]. BEBS future enhancements will aim to increase the metrics been calibrated based in use cases, and consideritng the type of communications to be exchanged.

The added value of this type of networks seems to be subtle [24]. The added value is what makes these networks an interesting opportunity for alternative business models based on individual allocation of access to more data. It is able to view more data on the user, some of which may be intentionally obscured from the public or strangers. This allows users to network with a specific user in a more intimate and personal setting. Furthermore, it facilitates the creation of greater communication options, which, depending on the social site, opens up new avenues of communication. This adds a greater level of interactivity: you can connect with the person who added you through private/direct messages, instead of the highly visible public channel.

Since users are able to recommended content, when someone adds someone else as a friend (and vice versa), activity or actions on the site may be recommended or 'pushed' towards the other user in some part of their administrative panel or profile. This means that users achieve greater automatic visibility whenever they use the social networking tool. And finally there is a great social verification, emerging from the auxiliary advantage of having many fans on social media networking tools. This social verification arises especially when there is some kind of self-ranking from the users according to the number of followers/subscribers using the tool. Popular and visible users tend to accumulate friends more easily than unknown users, and potentially be hubs for data distribution.

Overall what is more important is the fact that there is a value to this trading that can, with calibration estimates, make a strong case for the implementation of applications and systems in which the sustainability of alternative business models can be assessed, and their viability and economic profitability verified, since the investment for establishing such networks is practically null from the point of view of traditional telecoms.

After releasing the software JNLP application on the project's website (www.bionets.eu), the source code has been made available on Sourceforge. Further testing is currently been completed.

References

1. Agar, J.: Constant Touch: A Global History of the Mobile Phone. Icons Book Ltd., London (2004)
2. Barabasi, A.L.: Linked: The New Science of Networks. Perseus Publishing Books, Cambridge (2002)
3. Baka, V., Scott, S.V.: From Studying Communities To Focusing On Temporary Collectives: Research-In-Progress on Web 2.0 in the Travel Sector. Information Systems Working Papers. London, Department of Management, London School of Economics and Political Science: 12 (2008)
4. Bala, V., Goyal, S.: Learning from neighbours. Review of Economic Studies 65(3), 595–621 (1998)
5. Benkler, Y.: Sharing nicely. The Yale Law Journal 114, 273–358 (2005)
6. Benkler, Y.: The Wealth of Networks: How Social Production Transforms Markets and Freedom. Yale University Press, New Haven (2006)
7. BIONETS, Multiple authors Application Scenarios – Round II. Internal Document (2007)
8. BIONETS, Multiple authors ID3.3.2 Economics for BIONETS Business Models (2007)
9. Frogloop (2009, 2007). Is it Worth it? An ROI Calculator for Social Network Campaigns." Retrieved (February 2009), http://www.frogloop.com/care2blog/2007/7/17/is-it-worth-it-an-roi-calculator-for-social-network-campaign.html

10. Goldbeck, J., Hendler, J.: Inferring Trust Relationships in Web-Based Social Networks. ACM Transaction on Internet Technology 6(4) (2006)
11. Granovetter, M.: The Strength of Weak Ties: A network theory revisited. In: Marsden, P.V., Lin, N. (eds.) Social Structure and Network Analysis, Beverly Hills, USA, pp. 105–130. SAGE, Thousand Oaks (1982)
12. Harwood, R. (2008, December 02, 2008). Nesta Connect: Connecting Dots and Valuing Networks, `http://blogs.nesta.org.uk/connect/2008/12/connecting-dots-and-valuing-networks.html` (retrieved January 17, 2009)
13. Jelasity, M.: Engineering emergence through gossip. In: Proceedings of the Joint Symposium on Socially-Inspired Computing Hatfield, UK, University of Hertfordshire (2005)
14. Gonzales, M.C., Babarasi, A.-L.: Complex networks - From data to Models. Nature Physics 3 (2009)
15. Law, A.M., David, K.W.: Simulation Modeling & Analysis, 3rd edn. McGraw Hill International Editions, Singapore (2000)
16. Maki. The Value of Friends in Social Media Websites (2009),
`http://www.doshdosh.com/`
`the-value-of-friends-in-social-media-websites/`
(retrieved January 2009)
17. Priscilla. Does effort = effect? (2007),
`http://www.solidariti.com/article/Doesefforteffect/`
(retrieved February 2009)
18. Quiggin, J.: Why do social networks work? (2006),
`http://crookedtimber.org/2006/05/60/`
`why-do-social-networks-work` (retrieved January 2009)
19. Rheingold, H.: Smart Mobs. Perseus Publishing, Cambridge (2002)
20. Schrock, A.: Examining social media usage: Technology clusters and social network site membership. First Monday 14 (2009)
21. Schumpeter, J.A.: The economy as a whole - seventh chapter of the theory of economic development. Industry and Innovation 9(1/2) (2002)
22. Shirky, C.: Here Comes Everybody: The Power of Organizing Without Organizations. Allen Lanes for Penguin Books, London (2008)
23. Twitter. Twitter Value (2009) (retrieved February 2009), `http://tweetvalue.com`
24. Van Buskirk, E.: 5 Ways the Cellphone Will Change How You Listen to Music (January 16, 2009),
`http://blog.wired.com/business/2009/01/six-ways-cellph.html`
(retrieved January 2009)
25. Weinberger, D.: Everything is Miscellaneous - The Power of the New Digital Disorder. Times Books, New York (2007)
26. Wiredset. Types of Engagement (2006),
`http://wiredset.com/archives/2006/11/20_engagement.html`
(retrieved February 2009)

BIONETS: Self Evolving Services in Opportunistic Networking Environments

Iacopo Carreras[1], Louay Bassbouss[2], David Linner[2], Heiko Pfeffer[2], Vilmos Simon[3], Endre Varga[3], Daniel Schreckling[4], Jyrki Huusko[5], and Helena Rivas[5]

[1] CREATE-NET, Trento, Italy
[2] TUB, Berlin, Germany
[3] BUTE, Budapest, Hungary
[4] Univ. of Passau, Passau, Germany
[5] VTT, Finland

Abstract. This paper presents the *BIONETS* opportunistic service evolution platform. The proposed platform allows pervasive services to evolve over time by exploiting opportunistic communications among mobile nodes on the one hand, and evolutionary computation techniques on the other. We present the main components of the platform, describing their functionalities and technical implementation. Finally, we present the hardware–in–the–loop approach we have followed to evaluate it, where a simulation platform, in charge or reproducing a large number of mobile nodes communicating wirelessly, is integrated with a real software prototype.

Keywords: evolutionary services, opportunistic networking, hardware–in–the–loop, demonstrator, BIONETS.

1 Introduction

Opportunistic communication systems [1] have gained a significant attention from the research community. This is mostly due to the proliferation of mobile devices such as smartphones, equipped with short-range wireless connectivity (e.g., Bluetooth and WiFi) that have encouraged the development of applications which allow users to produce, access and share digital resources without the support of a fixed infrastructure. This includes not only digital content, but also mobile pervasive services residing on users portable devices. In particular, services become now able to interact with each other simply as the consequence of users co-location. This is enabling innovative execution models where services are able to share/exchange components, data and evolve over time in order to adapt their behavior to a specific situation or context.

Starting from this socio-technological trend, the *BIONETS* project [2] developed the concept of *opportunistic evolutionary services*, which consists in pervasive services evolving as the consequence of *P2P* localized interactions among mobile nodes. The reference scenario is constituted by mobile services, running on users' portable devices, and able to evolve as the result of (i) service execution by users (ii) interactions among different services running on various user devices. Evolutionary principles are applied to different generations of *composite services*, which are complex services resulting by

E. Altman et al. (Eds.): Bionetics 2009, LNICST 39, pp. 88–94, 2010.

the interwinding of atomic ones. The evolution process takes place by first evaluating the fitness of a given service composition, and then applying genetic operators to create new ones.

In this paper we will describe the BIONETS platform, which is the combination of an opportunistic content distribution framework and a distributed service execution engine based on the concept of portable services. The platform has been evaluated following a hardware–in–the–loop approach, where a simulation platform reproducing mobile nodes moving and exchanging data is integrated with a prototype of the application being provided to users.

The reminder of this paper is organized as follows. In Sec. 2 we will briefly describe the opportunistic service evolution concept. In Sec. 3 we will describe the demonstrated platform and the hardware–in–the–loop approach. Finally, in Sec. 4 we will conclude the paper, pointing out current research activities.

2 Evolutionary Opportunistic Services

The reference application scenario of the BIONETS project [2] is that of future computing environments: smart ambients characterized by a halo of heterogeneous mobile devices embedded in the environment and by mobile nodes exchanging data through localized interactions whenever in close proximity. Due to the mobility of nodes, disconnected operations represent the rule, rather than the exception, and information is diffused similarly to the spreading of an epidemic. *Opportunistic evolutionary services* refer then to mobile services being executed over such an opportunistic networking infrastructure, and evolving over time as the consequence of a distributed evolutionary process.

2.1 Epidemic Data Spreading

Starting from the considered application scenario, the BIONETS project developed the "disappearing networking" concept, which is a networking framework addressing the problems of scale (in terms of number of devices) and heterogeneity (in terms of different features supported by the different nodes). In particular, it provides a novel network architecture, centered around the concept of "epidemic spreading" of information: similarly to the spreading of an epidemic, data is diffused by means of localized interactions among mobile nodes. Data exchanges are regulated by a dissemination scheme, which determines the rules according to which data is forwarded from one node to another. Refer to [3] for a comprehensive overview of the different data forwarding schemes that can be used in order to deliver messages in an opportunistic communication environment. We have then designed and evaluated various data dissemination schemes, and studied how different message forwarding algorithms can co-exist on the basis of natural selection principles [4].

Various security mechanisms were also investigated to ensure classical security characteristics in such a non-classical environment. In particular availability, reliability, and trustworthiness of the information spread are critical aspects to be considered. For this purpose, we developed a so called *barter-based-approach* [5] which has the characteristic that the only beneficial behaviour for an attacker, such as a selfish-node, is beneficial for all nodes in the network.

This mechanism can be complemented by a *fair exchange protocol* which also allows the fair exchange of valueable information even though no central trusted third party is present. Here again, the characteristic of the disappearing network is exploited by defining a transient trusted third party formed by the surrounding nodes. Finally, the quality (or trustworthiness) of the information spread in BIONETS is evaluated by a trust and reputation management system. This is done by assessing the trust of the node the information originates from.

2.2   Evolutionary Mobile Services

Evolutionary services are expected to run on top of such "disappearing networking" infrastructure, and to be subjected to a specific service life-cycle which describes their creation, lifetime and possible deprecation. Traditional service life-cycles are rather static and do not possess abilities to dynamically respond to environmental changes. Differently, in BIONETS we are focusing on highly dynamic environments, which required the realization of a novel bio-inspired life-cycle, which features service compositions that can evolve over time to remain adapted to the environmental context and can support the security requirements therein [6].

The bio-inspired service life-cycle has been realized on top of the BIONETS *SerWorks* framework [7], which is depicted in Fig. 1. It consists of four loosely coupled containers. An *Application Container* holds atomic services as well as service compositions; a *Network Container* encompasses services providing networking functionalities. Both types of services are not extended to provide a certain autonomic behavior. Instead, a third container contains a set of *Mediators*, which realize basic functionalities such as service discovery/recovery, as well as complex operations such as the modification

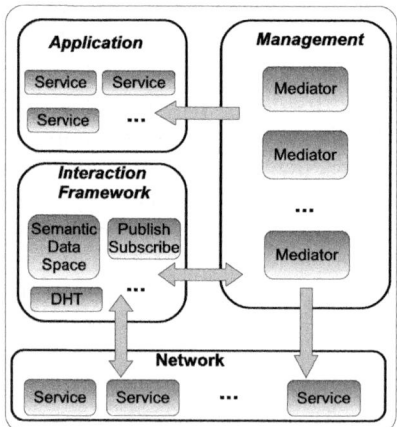

Fig. 1. The BIONETS Serworks system architecture

of service compositions by means of genetic operators. Interactions among elements of these containers are handled by the Interaction Framework, which implements various interaction models. As an example, in Fig. 1 a DHT, Puslish/Subscribe and Semantic Data Space interaction models are considered.

The interaction model includes also the support for service migration, a key element to support service evolution and to increase the service penetration in a dynamic environment. Service migration can be interpreted as a special case of a wider service mobility concept, where the running service, or set of services, are transferred from one node to another, while keeping all the user and network associations alive during the procedure. In the BIONETS platform, the service mobility and migration is handled by a specific software entity, which makes the decision and triggers the movement of the service over the existing network. In order to realize such procedure, such component needs to gather information also from the other containers (including e.g. the networks status information from the network container) and to interact with the other Mediators.

In this paper's scope, the BIONETS service migration is defined as node's capability to exchange service descriptions with other nodes over the network and within the connectivity range, and to execute the service with either the service's original preference set or if necessary with modified preference set corresponding the special requirements of the new platform, such as new user credentials etc. In this migration scenario, we do not assume multi-hop network and the migration happens over the point-to-point link.

Evolving Service Compositions. One of key objectives of the BIONETS project is to define innovative solutions for enabling the adaptation of service compositions during runtime. The many existing solutions are based on the simple replacement of services based on certain additional information. For instance, in [8] genetic operators are used to enable a QoS-aware creation of service compositions by selecting the single services within the composition based on their quality. However, they did not revise the structure of the service composition itself and thus did not address the problem of related semantic service descriptions. In [9], authors proposed a policy specification language and respective hierarchical policy model to incorporate the users context during the binding phase of a service compositions creation and thereby also provide a solution to adapt services based on a changing user context. Again, this solution focuses on an optimal replacement of single services, but cannot provide an alternative solution in case the service cannot be replaced by exactly one other service.

In BIONETS, we aim to overcome the limitations of information driven adaptation and replacement algorithms towards an autonomic evolution of applications, where underlying service compositions are continuously evolving over time in order to fit into the continuously changing environment. This is achieved through a service composition model where a workflow graph based on modified timed automata and a dataflow graph serve as the basis for evolutionary applications [10]. By applying genetic operators we are able to evolve service compositions in order to substitute parts of the application that have dropped out and to provide an equivalent functionality with a different set of services [11]. Equivalently, we developed mechanisms [12] which modify the service composition model in order to guarantee security requirements of the user, of the user's data the service uses to operate, or the device the service is running on.

3   The BIONETS Platform Demonstrator

The BIONETS platform is a practical implementation of a mobile framework supporting the execution of *evolutionary opportunistic services*. It consists of (i) a software
prototype of the platform (ii) a simulation environment, which is used to reproduce a
very large number of mobile nodes exchanging data over time. The software prototype
and the simulator are integrated following a *hardware–in–the–loop* approach, where the
scale of a very large number of mobile nodes is delegated to a simulation environment,
and the user interactions of the developed application is implemented in a real prototype
(Fig. 2). More in detail, this consists of a server simulating a mobile network containing
a large number of virtual nodes. The prototype connects to this server and participates
in the simulated network, just like it was a real one. This allows to stress test the prototype in a much larger environment that is not practically possible by using real devices.
In this setting, the prototype is represented in the simulation through a *delegate* node,
which is a simulated node implementing all the networking functions and interacting
with the Service Framework that is running in the prototype.

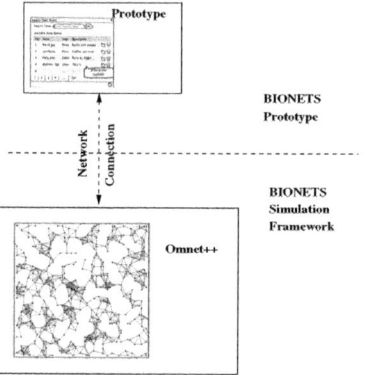

Fig. 2. The hardware–in–the–loop demonstrator architecture

3.1   Simulating an Opportunistic Network

To verify and benchmark the system properties of BIONETS, a special purpose simulation platform was implemented. This was based on $OMNeT++$ [13], a C++ simulation library which includes the support for simulating a mobile network. The focus of
the simulator is the implementation of the networking functions of the SerWorks architecture Fig. 1, which specifies a service oriented dynamically configurable networking
solution. This module encapsulates all of the opportunistic networking and epidemic
spreading related protocols, as well as many of the security solutions (e.g., trust and
reputation management, fair exchange and barter based protocols). The challenges related to the development of this platform mostly relate to the (i) implementation of a
platform that was able to scale up to a very large number of nodes (ii) support of a wide
range of different data dissemination schemes that could be selected at runtime.

3.2 Evolving Service Compositions

In BIONETS, composed applications are modeled by means of service compositions that are represented as timed automata [14]. This formal notion constitutes a formal link to mature research areas such as semi-group or category theory, making existent algorithms applicable to the BIONETS service composition notion. In addition, the special consideration of real-time aspects makes the composed application responsive to timeout and delays, such that services can by dynamically exchanged during runtime in case a service or its host device is no longer responding.

A runtime environment has been developed that supports the execution of these timed automata based service compositions. An implementation is available both for services as well as for smaller mobile devices. The latter version is implemented in pure JavaScript, such that it can be initially transferred to the client device and run on every user device featuring a Web browser with a JavaScript interpreter; changes on client devices are thus not necessary.

Services are labeled with semantic descriptions specifying their inputs, outputs, pre-conditions, and effects, so-called IOPE descriptions [15]. Based on these annotations, algorithms have been created to automatically create a service composition for a requested set of effects and outputs. Within the scope of our demonstration environment, variants of these algorithms are used to dynamically create service compositions and to evolve them over time in order to remain adapted to the continuously changing environmental context. Here, a user can initially subscribe to a set of services or functions. Appropriate services are discovered and offered to the user. In case a service is considered as suitable, it is automatically integrated into the user's current service composition. Every time new services become available within the network, services are offered to the user and optionally integrated during runtime. Thereby, the application evolves over time, offering an adapted application to the user given the current landscape of available services. To achieve a mapping from service compositions underlying evolvable Web applications, we developed a mapping from the service compositions to a graphical presentation. Here, services are classified according to the resources they generate. For instance, there are services operating on map, others on locations, and others on images. The resource allows us -in the style of the Web- to derive a representation of the current state of the service composition. Every time a resource is modified by means of a service execution, the accordingly modified resource is pushed to the frontend such that its presentation can be updated.

4   Current Research Work

In this paper we have introduced the concept of evolving opportunistic services, we have described their practical implementation over the BIONETS platform. Current research activities are currently concerned with the implementation of *MyDirector*, an application that loosely combines mobile services for the presentation of location-based data. The application supports the geo-referenced capturing and sharing (via opportunistic networking) of video snapshots in an urban environment. This use-case will be used as a driver for the evaluation of opportunistic evolvable services in a concrete application scenario, providing a valuable feedback on the developed platform and a benchmark for the implemented algorithms and solutions.

Acknoledgements

This work has been supported by the European Commission within the framework of the BIONETS project ISTFET-SAC-FP6-027748, www.bionets.eu.

References

1. Pelusi, L., Passarella, A., Conti, M.: Opportunistic networking: data forwarding in disconnected mobile ad hoc networks. IEEE Comm. Mag. 44(11) (November 2006)
2. BIONETS: Bio-inspired evolution for the pervasive age, http://www.bionets.eu
3. Hanbali, A.A., Ibrahim, M., Simon, V., Varga, E., Carreras, I.: A survey of message delivery protocols in mobile ad hoc networks. In: Proc. of INTERPERF, Athens, Greece (October 2008)
4. Simon, V., Bérces, M., Varga, E., Bacsardi, L.: Natural selection of message forwarding algorithms in multihop wireless networks. In: Proc. of IEEE WiOpt, Seoul, Korea (June 2009)
5. Buttyan, L., Dora, L., Felegyhazi, M., Vajda, I.: Barter-based cooperation in delay-tolerant personal wireless networks. In: Proc. of WoWMoM (2007)
6. Pfeffer, H., Linner, D., Radusch, I., Steglich, S.: The bio-inspired service life-cycle: An overview. In: Proc. of ICAS, Athens, Greece (June 2007)
7. Pellegrini, F.D., Miorandi, D., Linner, D., Bacszardi, L., Moiso, C.: Bionets: from networks to serworks. In: Proc. of SAC, Budapest, Hungary (December 2007)
8. Canfora, G., Penta, M.D., Esposito, R., Villani, M.L.: An approach for qos-aware service composition based on genetic algorithms. In: Proc. of ACM GECCO, New York, NY, USA (July 2005)
9. Zhang, B., Shi, Y., Xiao, X.: A policy-driven service composition method for adaptation in pervasive computing environment (2008)
10. Pfeffer, H., Linner, D., Steglich, S.: Modeling and controlling dynamic service compositions. In: Proc. of IEEE ICCGI, Washington, DC, USA (August 2008)
11. Linner, D., Pfeffer, H., Steglich, S.: A genetic algorithm for the adaptation of service compositions. In: Proc. of SAC, Budapest, Hungary (December 2007)
12. Schreckling, D.: Adaptive Security in BIONETS. BIONETS (IST-2004-2.3.4 FP6-027748), Deliverable D4.4 (Feburary 2009)
13. OMNeT++: Discrete event simulation system, http://www.omnetpp.org
14. Pfeffer, H., Linner, D., Steglich, S.: Modeling and controlling dynamic service compositions. In: Proc. of ICCGI, Washington, DC, US (July 2008)
15. Jaeger, M., Engel, L., Geihs, K.: A methodology for developing owl-s descriptions. In: Proc. of INTEROP-ESA, Geneva, Switzerland (February 2005)

Activation–Inhibition–Based Data Highways for Wireless Sensor Networks⋆

Daniele Miorandi[1], David Lowe[1,2], and Karina Mabell Gomez[1]

[1] CREATE-NET, v. alla Cascata 56/D, 38123 – Povo, Trento, IT
{daniele.miorandi,david.lowe,karina.gomez}@create-net.org
[2] Centre for Real-Time Information Networks, UTS, Ultimo, NSW 2007, Australia
david.lowe@uts.edu.au

Abstract. In this work we report on a method, based on the use of activation–inhibition mechanisms, for building and maintaining high–rate routing paths (data highways) in dense wireless sensor networks. We describe the algorithms devised and report on the outcomes of a simulation study, aimed at assessing the scalability and the performance of the proposed approach.

Keywords: activation–inhibition mechanisms, reaction–diffusion patterns, data highways, routing, wireless sensor networks.

1 Introduction

The design of efficient, robust and scalable methods for routing data from sources to sink(s) is a major issue in wireless sensor networks (WSNs) [1]. From a communication perspective, traffic patterns in WSNs are of the many–to–one or many–to–some type, depending on the presence of one or multiple data sinks. Data sinks can be either gateway nodes, through which the sensed information, appropriately processed, can be accessed by remote machines, or actuators (e.g., PLCs), where control decisions are taken based on the physical phenomena monitored by the WSN.

Inspired by the work of Franceschetti et al. [5] on the possibility of achieving the capacity bound in randomly deployed wireless ad hoc networks by using a set of high–rate wireless backbones or 'highways' spanning a given network strip, we introduced in a companion work [6] a self–organizing scheme for the distributed construction of data highways in dense WSNs. The scheme is based on the use of activation–inhibition mechanisms [3] for driving the emergence of suitable *spatial patterns*. Similar approaches have recently found application in the wireless networking setting for dealing with activation problems [4,8]. The scheme presented in [6] assumes that all nodes have perfect information on the relative location of data sinks. Such information, which is used to "orient"

⋆ This work has been partially supported by the European Commission within the framework of the BIONETS project EU-IST-FET-SAC-FP6-027748, www.bionets.eu

E. Altman et al. (Eds.): Bionetics 2009, LNICST 39, pp. 95–102, 2010.

the activation–inhibition filter along the node–to–sink direction, may not be available in realistic deployments. We therefore investigated a refined version of such a model, where a beaconing mechanism is used to allow each node to estimate the direction towards the sink(s) (i.e. which neighbouring nodes lie on the minimum hop path towards the sink). In this work, we describe the refined model and present some early–stage results derived from an implementation in an event–driven simulator of the aforementioned techniques.

The remainder of the paper is organized as follows. In Sec. 2 we introduce the architecture and the methods for the construction and maintenance of data highways. In Sec. 3 we present the outcome of our simulative study. Sec. 4 concludes the paper presenting a brief overview of the issues left open.

2   Architecture and Methods

2.1   Highway Generation

We wish to develop self-organizing processes that lead to the emergence of data highways in a dense wireless network. These highways should be optimally spaced such that all nodes are within range of a highway (using long–range single–hop communications), but the highways themselves should utilize short–range hops to transport messages to data sinks (to optimize power consumption while limiting interference). The goal is to minimise the number of highway nodes while respecting such constraints. Our problem is therefore to approximate such a desired pattern by using decentralised mechanisms which exploit only local interactions among neighbouring nodes.

We took inspiration, for addressing such a problem, from mechanisms based on activation–inhibition reaction–diffusion techniques [2,8]. These mechanisms describe how field strengths or substance concentrations vary over space and time under two competing influences, a short range positive activation effect, and a longer range negative inhibition effect. The resultant models have been widely used to describe emergent behaviours in biological and physical processes [3]. The simplest formulation of this approach, yet enabling the emergence of a variety of patterns of interest, makes use of a single field variable and can be modelled in the discrete time domain as follows:

$$u(\mathbf{k}, t+1) = g\left[\varphi_s u(\mathbf{k}, t) + \sum_{\mathbf{j} \in R_i} \varphi_i(\mathbf{j}) u(\mathbf{k}+\mathbf{j}, t) + \sum_{\mathbf{j} \in R_a} \varphi_a(\mathbf{j}) u(\mathbf{k}+\mathbf{j}, t)\right], \quad (1)$$

where $\mathbf{k}$ is a physical location, R_i is the inhibition region, R_a the activation region, the activation coefficient φ_a and the self–activation coefficient φ_s are strictly positive, the inhibition coefficient φ_i is strictly negative and $g()$ is a normalizing function.

The equation in (1) corresponds to a two–dimensional time–invariant filter applied to the random field representing the activation level $u(\cdot, \cdot)$. The filter is characterized by the shapes of activation and inhibition regions and by the value of the coefficients $\varphi_s, \varphi_a, \varphi_i$. In [6] it has been shown that, by using asymmetric (or polarised) filters (i.e., filters in which the inhibition and activation regions are not symmetric) it is possible to achieve various spatial patterns. The repeated filter convolution causes the emergence of the ridge peaks in the activation field by activating localized regions that align with the filter axis, whilst inhibiting the off–axis areas between these regions. The width of the filter's inhibition zone controls the separation of the resultant ridge peaks. It is therefore possible to select filter parameters to achieve desired ridge separations. By appropriately adapting the direction of the axis throughout the network, it is possible to change the ridge orientation in different locations within the network. If this is done appropriately then the ridges can be controlled to converge on specific locations – i.e. the data sinks within the network. If we have multiple data sinks then the filter orientation, and hence the ridge orientation, can be derived based on a gravitational attraction model [6]:

$$
\mathbf{d_i} = \frac{\sum_{s_j \in S}(\mathbf{n_i} - \mathbf{s_j})|\mathbf{n_i} - \mathbf{s_j}|^{-2}}{\sum_{s_j \in S}|\mathbf{n_i} - \mathbf{s_j}|^{-2}}, \tag{2}
$$

where $\mathbf{S}$ is the set of sink nodes.

Such a method is based on the assumption that nodes are able to estimate the direction to all data sinks. In the absence of such information, nodes have to estimate which nodes, among the neighbouring ones, are along the shortest path to the sink. This can be achieved by using a suitable beaconing mechanism to acquire knowledge of the distance from the sink(s).

As we are assuming that the nodes *do not* know their own spatial location nor that of the sinks, the convolution filter cannot be oriented based on equation (2). We can however obtain an estimate of the direction to the sink from a given source node based on the sink hop count of the neighbouring nodes, and in particular whether or not they are on the shortest path to the sink. If the neighbourhood has size R (expressed in hops) and the selected node is at distance δ from a given sink, the activation region is constituted by all nodes being at distance n from the node ($n \le R$) and at distance $\delta \pm n$ from the sink. All other nodes in the neighbourhood are considered to be in the inhibition region. One additional implication is that —rather than determining the weighted direction to all sinks as in (2)— given that each node knows the distance to all sinks, we apply the filter convolution separately for each sink, and hence determine a set of independent ridges for each sink. The highways are then only generated at each node using the ridge data for the closest sink.

2.2   Protocols Description

Formally, we assume the following:

- Nodes are assigned a unique identifier;
- Nodes can tune dynamically their transmission power level P_{tx} in the range $[P_{min}, P_{max}]$;

- The network is connected when all nodes use $P_{tx} = P_{min}$;
- Nodes transmit at P_{min} unless otherwise specified.

The algorithm works according to the following steps:

(i) **Sink announcement**: each sink broadcasts a beacon with its ID. Nodes receiving such a beacon update the `distance` field and re-broadcast it. In such a way each node will eventually have knowledge of all sinks and its distance from them.

(ii) **Neighbourhood discovery**: each node broadcasts a message to its one–hop neighbours, asking for information about them (including activation state) and about nodes which are at most $R - 1$ hops from them. In such a way, each node may acquire (by means of a gossiping mechanism) information on nodes that are within its 'neighbourhood' (at most R hops away).

(iii) **Filter construction and activation level update**: based on the information gathered, each node constructs its local filter and updates its activation level. Let us now consider the specific algorithm for performing the activation–inhibition convolution within each node. If a neighbouring node is n hops from the selected node ($n \leq R$), and either n hops closer to, or n hops further away from, the sink node, then the neighbour will be on the shortest path to/from the sink - and hence can be viewed as being along the filter alignment axis, and should therefore be an activator for the selected node. Conversely, a node that is n hops from the selected node, and not n hops closer to, or n hops further away from sink, will not be on the shortest path, and can therefore be considered to be orthogonal to the filter axis, and will therefore be an inhibitor for the selected node. *Steps (ii) and (iii) are repeated for a number of times in order to achieve a stable spatial pattern.*

(iv) **Ridge detection and tracing**: each node runs a procedure for deciding whether it should act as highway node, by checking whether it is in range of an already existing highway (in which case it just connects to it) or whether it is a local peak, and should therefore activate (i.e., turn into a highway node). If a node is activated, it starts a tracing process by identifying which node shall represent the next hop on the way to the closest sink.

The detailed description of the algorithms, together with their pseudo-code, are reported in [7].

3   Numerical Results

All the algorithms and protocols devised have been implemented in an event–driven network simulator, Omnet++ [9]. The model used was based on the IEEE 802.15.4 standard for PHY and MAC protocols. Our primary goal was to understand how the methods introduced scale with respect to the number of nodes in the network. For scaling the system in a consistent way, we used a fixed node density (set to 0.1429 nodes/m^2, corresponding to an average neighbourhood size equal to 7 when transmitting at the minimum power). The larger the number of nodes, the larger the playground size on which they were placed. Nodes were placed according to a uniform distribution. We used one single sink, located at $(1, 1)$. The maximum and minimum communication distance were setup considering, in the absence of interference, a power transmission range of $(-25, 0)$ dBm, a signal attenuation threshold of -120 dBm and a path loss coefficient of 2. The following set of parameters were used:

Parameter	Value
Number of runs	10
Duration of each run (s)	3600
Playground area (m^2)	70/350/700/1400/3500/5250/7000/10500
Number Hosts(n)	10/50/100/200/500/750/1000/1500
Maximum communication distance (m)	12.0
Minimum communication distance (m)	4.0
Neighbourhood size $(hops)$	4
Self-activation coefficient(φ_s)	2.0
Activation coefficient(φ_a)	1.0
Inhibition coefficient(φ_i)	-0.1

In all simulation runs our protocol showed to be able to build valid routes, i.e., all nodes had a valid nextHop field, highways were connected to the sink and nodes not on highways were within the maximum communication distance of 12 m from a highway node.

In order to provide insight into the patterns arising in the network as a consequence of the activator–inhibitor mechanism, we report in Fig. 1 the following graphs, related to the case with 750 nodes:

(a) Connectivity graph among nodes when transmitting at minimum transmission power;
(b) Contour plot of resulting distance from the sink;
(c) Resulting activation field after 25 iterations;
(d) Data highways resulting from tracing activation field ridges to the sink.

As it can be seen, highways tend to arise where dense clusters of nodes are present; the ridge tracing process ensures then ability to reach back to the sink. The control of the distance among highways is achieved by setting appropriately selecting the neighbourhood size for the activation filter.

In order to evaluate the performance of our protocol, we preliminarily focused on two metrics. The first one is the time needed by a node in the network to achieve a valid path to the sink node. This corresponds to the time needed to bootstrap a WSN. We considered the minimum, average and maximum value attained for any node over 10 runs. The results are reported in Fig. 2. Both the minimum and average number turn out to be only slightly sensitive to the number of nodes in the system. This is due to the fact that the time needed to construct routing paths from any node to the sink turns out to be dominated by the values of some timers which are part of the routing framework (see Fig. 4 for a workflow–like representation of the overall operations of the protocol). The case of 10 nodes show significantly better performance, due to the simple topology achieved (in most runs all nodes were directly connected to the sink).

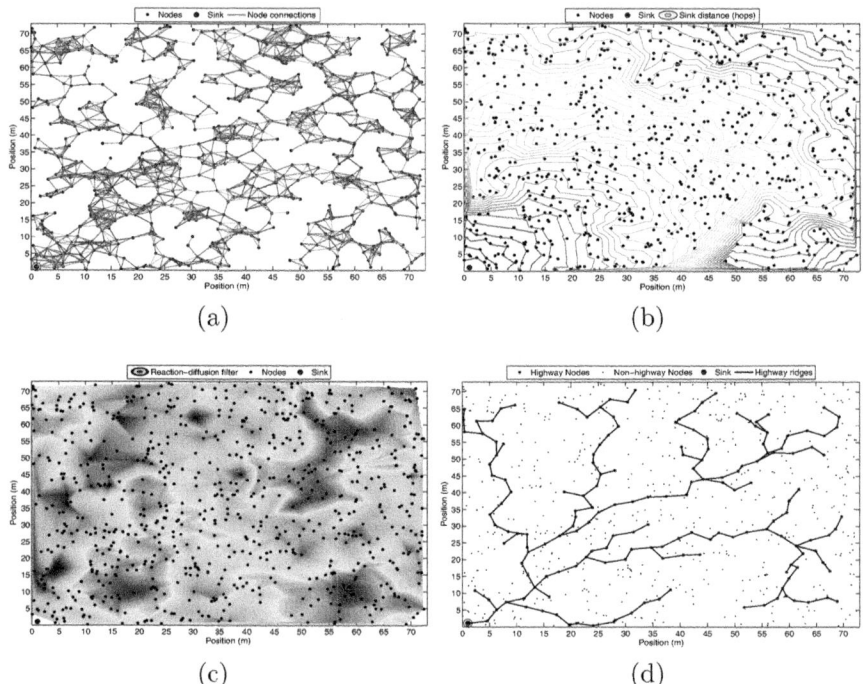

Fig. 1. Pattern arising for $N = 750$: (a) Connectivity graph when transmitting at the minimum power level; (b) Contour plot of distance from the sink; (c) Activation field after 25 iterations of the reaction–diffusion filter; (e) Data highways resulting from tracing activation field ridges

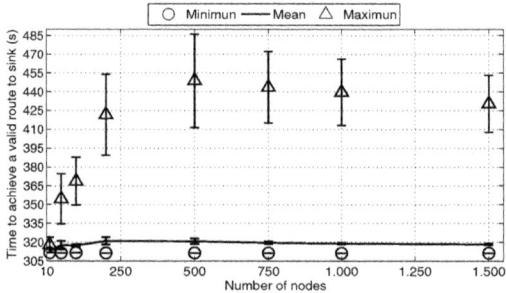

Fig. 2. Bootstrapping time (s) as a function of the network size

The maximum value increased as a function of the number of nodes. A more detailed analysis revealed that this was due to problems related to interference, which prevented some nodes to correctly decode messages destined to them, causing therefore a delay in the setup time.

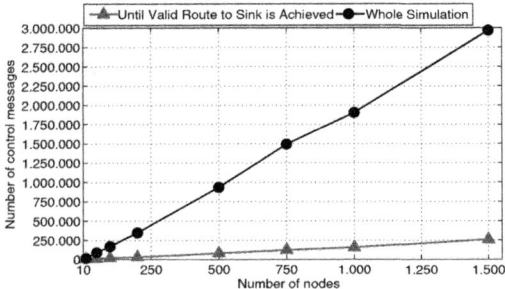

Fig. 3. Number of control messages (exchanged throughout the whole simulation and until a valid route to sink is achieved) as a function of the network size.

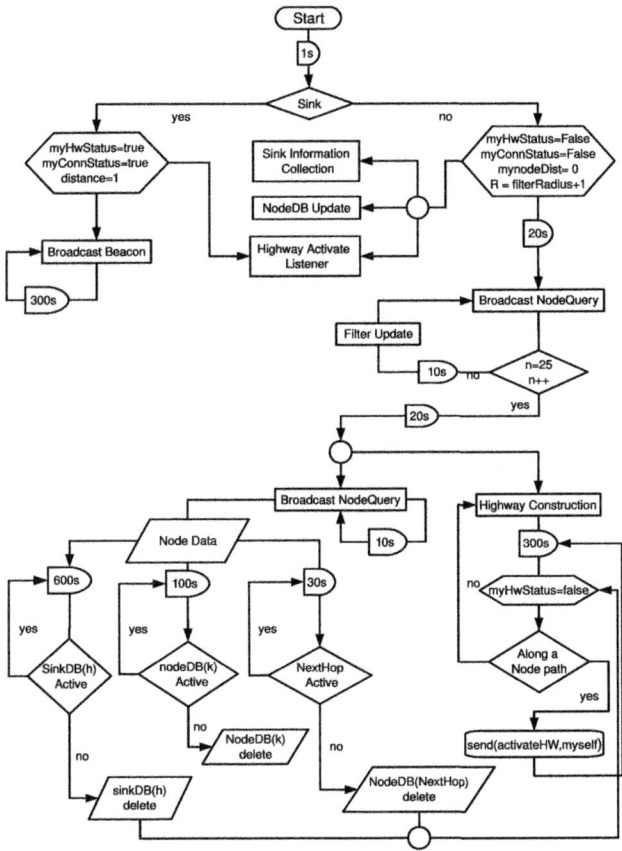

Fig. 4. Workflow–like representation of the operations of the whole routing framework. (Values for duration of timeouts are indicative and have been derived for the network sizes considered in this paper.)

The second performance metric we considered was related to the overhead induced by the protocol in terms of the number of messages exchanged until a valid route to sink is achieved and throughout the whole simulation run (of the duration of 3600 s). As it may be seen from Fig. 3, the number of such a messages scales (slightly) superlinearly in the number of nodes. A more detailed analysis, based on the different types of messages involved in the routing protocol, revealed that all messages involved but sink beacons grow linearly with the number of nodes. The number of sink beacon messages tends on the other hand to grow in a superlinear fashion (actually exponentially), which explains the behaviour observed in our experiments.

4   Conclusions and Discussion

In this paper, we have reported some early–stage numerical results on the performance attainable by a biologically–inspired scheme for the construction of high–rate data highways in dense wireless sensor networks. The results confirmed (i) the correctness of the procedures designed and their ability to meet the requirements in terms of maximal distance to highways (ii) good scalability properties with respect to the number of nodes in the system. Further simulation studies are needed to assess the performance attainable in terms of energy consumption as well as in the ability of the mechanisms designed to recover quickly and with limited overhead from nodes' failures.

References

1. Akyildiz, I.F., Su, W., Sankarasubramaniam, Y., Cayirci, E.: Wireless sensor networks: a survey. Computer Networks 38(4), 393–422 (2002)
2. Bar-Yam, Y.: Dynamics Of Complex Systems. Westview Press (2003)
3. Deutsch, A., Dormann, S.: Cellular automaton modeling of biological pattern formation: characterization, applications, and analysis. Birkhäuser, Basel (2005)
4. Durvy, M., Thiran, P.: Reaction-diffusion based transmission patterns for ad hoc networks. In: Proc. of IEEE INFOCOM, pp. 2195–2205 (2005)
5. Franceschetti, M., Dousse, O., Tse, D.N.C., Thiran, P.: Closing the gap in the capacity of wireless networks via percolation theory. IEEE Transactions on Information Theory 53(3), 1009–1018 (2007)
6. Lowe, D., Miorandi, D.: All roads lead to rome: Data highways for dense wireless sensor networks. In: Proc. of S-Cube 2009: The first international conference on Sensor Systems and Software, ICST (2009)
7. Miorandi, D., Lowe, D., Gomez, K.M.: Distributed bio-inspired methods for data highways construction and maintainance in WSNs. Technical Report 200900022, CREATE–NET (September 2009),
   http://www.create-net.org/~dmiorandi/TR200900022.pdf
8. Neglia, G., Reina, G.: Evaluating activator-inhibitor mechanisms for sensors coordination. In: Proc. of Bionetics, Budapest, Hungary, ICST (2007)
9. OMNeT++ discrete event simulation system, http://www.omnetpp.org

Minimum Expected *-cast Time in DTNs

Andreea Picu and Thrasyvoulos Spyropoulos

ETH Zürich, Switzerland
lastname@tik.ee.ethz.ch

Abstract. Delay Tolerant Networks (DTNs) are wireless networks in which end-to-end connectivity is sporadic. Routing in DTNs uses past connectivity information to predict future node meeting opportunities. Recent research efforts consider the use of social network analysis (i.e., node communities, centralities etc.) for this forecast. However, most of these works focus on unicast. We believe that group communication is the natural basis of most applications envisioned for DTNs. To this end, we study constrained *-cast (broad-, multi- and anycast) in DTNs. The constraint is on the number of copies of a message and the goal is to find the best relay nodes for those copies, that will provide a small delivery delay and a good coverage. After defining a solid probabilistic model for DTNs collecting social information, we prove a near-optimal policy for our constrained *-cast problems that minimizes the expected delivery delay. We verify it through simulation on both real and synthetic mobility traces.

Keywords: DTN, social network, broadcast, multicast, anycast.

1 Introduction

Delay tolerant networks (DTNs) are sporadically-connected networks, experiencing frequent network partitioning. This is usually the result of one or more factors: high mobility, low node density, stringent power management, attacks etc. The DTN concept was first considered for challenged or exotic network conditions such as satellite networks with periodic connectivity, underwater acoustic networks with moderate delays and frequent disruptions, sensor networks for wildlife tracking, and Internet provision to developing regions. Nonetheless, the constant growth in number of mobile, networking-enabled devices has created the possibility for anyone to set up a network anywhere, anytime and with anybody, even in urban environments, something that has not remained unnoticed by researchers and industry. Numerous applications, e.g., peer-to-peer content exchange, localized content and service discovery, social networking, that are asynchronous and thus tolerant to delays, can be supported using a DTN architecture.

Nevertheless, communication in such networks is quite challenging due to the lack (or rapid change) of end-to-end paths. To this end, numerous novel DTN routing protocols for unicast traffic [23, 9, 19, 17] have been proposed, that attempt to infer the contact patterns between nodes and predict paths with high probability of delivery. Among these, a recent research thread has proposed to take advantage of *social* properties of such networks [8, 16]. Mobile devices are carried by humans whose mobility is governed by explicit (e.g., friends, colleagues) or implicit social factors (e.g., commuters on the same bus, shared

E. Altman et al. (Eds.): Bionetics 2009, LNICST 39, pp. 103–116, 2010.
© Institute for Computer Sciences, Social-Informatics and Telecommunications Engineering 2010

cafeteria etc.). As a result, these protocols try to expressly capture social links between nodes (e.g., communities) and use complex network analysis tools [3, 22] to identify nodes that can be used to improve the routing efficiency.

Although unicast routing in DTNs has received a considerable amount of attention (see survey in [25]), other communication patterns, such as multicast or anycast, as well as appropriate metrics and algorithms to optimize these, have been largely neglected [13, 26]. However, as has been noted in [5], most applications envisioned for such "pocket switched networks" [15] are expected to not be point-to-point. For example, pushing data towards a subset of interested subscribers to a channel or service can be described better as multicast or broadcast, while service or data discovery that does not know (or care about) the identity of the node that provides the service in advance can be better described as anycast.

In this paper, we take a first step towards optimizing *-cast (i.e., broadcast, multicast, and anycast) in DTN environments of human carried wireless devices. The approach we take will also be based on social network analysis, as we believe that the preliminary advantages that have been demonstrated for unicast routing [8, 16] are well-grounded, independently of the communication mode studied. Our work has two main contributions. First, we develop a solid probabilistic model of predictive value for any DTN gathering social information, e.g., node degrees. Second, we prove a near-optimal policy for a class of *-cast problems in a resource constrained setting that minimizes the expected delivery delay.

The rest of the paper is structured as follows. Section 2 presents the network model used throughout the study and we formally define the problems we analyze. Section 3 comprises the probabilistic model and the proof of delay optimization. A preliminary evaluation is shown in Section 4. We conclude in Section 5.

2 DTN Network Model

In this section, we present the network model used in our study and we formally define the problems we subsequently address.

2.1 Network Model

Let $\mathcal{N}$ be the set of all nodes in the network, $|\mathcal{N}| = N$. Each of the N nodes is identified by a unique ID and its mobility is assumed to be governed by (implicit or explicit) social relations. Specifically, (i) we can identify *node communities*, i.e., sets of nodes that tend to meet each other preferentially, and ii) nodes have different *sociability* or number of nodes they meet in a given time interval, ranging from *solitary* to *gregarious*. This seems a reasonable assumption, since mobile devices (network nodes) are carried by humans, who engage in socially meaningful relationships. Several previous DTN experiments, [10, 18], have confirmed this type of interaction patterns and information flow. In this paper, we will focus on the latter characteristic only, and defer studying community structure for future work. Specifically, we assume a *social graph* is created using past *contacts* between nodes and we will optimize around the degree distribution of the graph.

Contacts. A *contact* between two nodes happens when those nodes have setup a bi-directional wireless link between them. We assume that:

i) contacts last for a negligible time compared to that between two successive
   contacts, but long enough to allow all the required data exchanges to happen,
ii) contacts occur in sequence, i.e., there are no simultaneous contacts[1].

Each contact has a defining feature: its *type*. The type of a contact is uniquely
defined by the IDs of the two nodes taking part in it. Hence, there are $\binom{N}{2}$
possible types of contact. We will assume that:

iii) the types of successive contacts are mutually independent random variables.

The distribution of each variable is fully defined by a $\binom{N}{2}$-sized vector of probabilities summing to 1. The probabilities depend on the mobility model and can be
estimated in function of the information assumed available. In our case, mobility
behavior is captured in a social graph, described next, and contact probabilities
depend on the node degrees in this graph (as shown in Section 3.1).

Social graph. A *social graph* represents our network: nodes (mobile devices)
are vertices and contacts are edges. The graph seeks to capture the aforesaid
social features of this network. Ways of creating the social graph of a DTN are
implicitly used in various previous works [8, 16]. More recently, [14] explicitly
addresses this as a standalone issue. Here, we build the social graph as follows.

Time is divided in W-sized windows. When a contact occurs, the respective
edge is added to the graph. The node degree shows how many **different** devices
that node contacted within one time window. Hence, the graph is simple: multiple contacts between the same two nodes result in only one edge. At the end of
each window, nodes update their running average degrees over all time windows.
This graph allows us to identify important actors in the underlying social network, by using complex network analysis metrics such as centrality. Following
the practice of [20], we make the further assumption that the maximum degree
in the graph is $o(\sqrt{N})$, reflecting the fact that in a large enough social network, a
single person, even a very social one, cannot know a constant fraction of all users.

2.2 Constrained *-Cast Problem

We now describe the scenarios we will address in our probabilistic analysis. All
scenarios use the above network model.

We focus on a subclass of constrained group communication problems in the
DTN environment: broadcast, multicast and anycast. The constraint under consideration is on the number of copies of a message. The problem is formally
defined below for the general case. Specific definitions for broadcast, multicast
and respectively, anycast follow.

Definition 1 (General case). *At time 0, a source node $s \in \mathcal{N}$, creates a mes-
sage m to be delivered to a fixed set $\mathcal{D}$ ($|\mathcal{D}| = D$) of distinct destination nodes in
the network, with $\mathcal{D} \in \wp(\mathcal{N})$ and $1 \leqslant D \leqslant N$. s must then find a set of nodes $\mathcal{L}$ of
size L which will each store a copy of m. The general goal is that every node in
$\mathcal{D}$ receive m. Maximizing the number of nodes that receive it before a deadline
(time to live (TTL)) or minimizing the delay until all nodes receive the message
are different flavors of the optimization problem in hand[2].*

---

[1] This is the case, for example, if we assume that the arrival process of contact events
  is Poisson. In general, this assumption just implies a relatively sparse network.

[2] $\wp(\mathcal{N})$ is the power set of $\mathcal{N}$, i.e., the set of all subsets of $\mathcal{N}$.

Definition 2 (Broadcast). *In broadcast, $\mathcal{D} = \mathcal{N}$.*

Definition 3 (Multicast). *In multicast, $\mathcal{D}$ is such that $2 \leqslant D \leqslant N - 1$.*

Definition 4 (Anycast). *In anycast, still $\mathcal{D} \in \mathcal{P}(\mathcal{N})$ and $1 \leqslant D \leqslant N$. However, the goal of the routing changes. Instead of showing the message to the entire destination set, m need now only be seen by at least one node belonging to that set, whichever node that is.*

Therefore, with our routing requirements, the creator of a message must find a set $\mathcal{L}$ of *permanent message carriers* and give them each a copy of the message. By permanent message carriers we mean that the carriers cannot forward the message further, except to nodes in the destination set $\mathcal{D}$ who consume it (similar to the two-hop [4] and spray and wait schemes [23]). The size L of the set of permanent message carriers should be at most equal to the permitted number of copies of a message. Since the copies cannot be forwarded, the choice of permanent message carriers must be optimum to ensure the timely delivery of the messages.

Our analysis focuses on the delivery delay measured in number of contacts. Let T_d be the random variable counting the delivery delay of a message under the above routing constraints. Then, T_d can be expressed as a sum of two other random variables: i) *the finding time*, the time for choosing the carriers and handing them the copies, T_f and ii) *the showing time*, the time until all destination nodes have seen the message, T_s. Assume[3] $T_d = T_f + T_s$. Based on this decomposition of the delivery delay, we can identify three levels of the problem:

1. **Offline optimization with global knowledge.** Here, we assume the availability of a *graph oracle* that provides information about the degree of all nodes in the social graph. Our goal is to select the L permanent message carriers, according to their (globally known) degrees, that will maximize the delivery ratio and minimize the delivery delay. Furthermore, we assume that after the set $\mathcal{L}$ is chosen, copies of the message are transmitted instantaneously to the nodes in this set. Hence, we ignore T_f and focus on T_s, the showing time.
2. **Online optimization with global knowledge.** We still have the same oracle. Using the scheme devised above, we identified the L permanent message carriers and our new goal is to find these carriers as soon as possible and give them the message. This part will assess only the T_f element.
3. **Online optimization.** We no longer have the oracle. Using insight from the two schemes above, we want to optimize the delivery delay T_d, in its entirety.

In this paper, we address the first level, i.e., we prove an optimum for $\mathbb{E}[T_s]$ and sharp concentration of T_s around its mean. The analysis of the next two levels is part of our future work.

3 Probabilistic Analysis

In this section we provide a theoretical analysis of the evolution of T_s, the showing time component of the delivery delay, in function of the choice of permanent message carriers, $\mathcal{L}$. To do so, let us first provide some basic tools.

---

[3] In reality, T_f and T_s will overlap. We discuss this later on.

3.1 Probability Measure

Consider the *contact*, as defined in Section 2.1 and let us place ourselves at time t. Define the random variable C_t as the type of the first contact occurring after time t in the network. This type is unique, since there are no simultaneous contacts and we will denote it (i, j), where i and j are the two nodes coming in contact.

As stated before, estimates of the probability distribution of C_t can be obtained, using the information available. In our case, that information is twofold. First, we know that the structure of our network is strongly influenced by social relations and second, we dispose of a degree distribution oracle. Several studies, [22, 3], found that most social networks have skewed degree distributions and nodes are linked proportionally to their degrees. For us, this suggests that the probability of C_t being (i, j) should be proportional to d_i and d_j, the degrees of i and j, respectively. We will now provide a formal argument that it is indeed true.

Let us now define our probability space $(\Omega, \mathcal{F}, \mathbb{P})$:

Ω is the sample space, i.e., the set of all outcomes of our elementary experiment, the next contact. Then, Ω consists of all possible pairs of nodes and $|\Omega| = \binom{N}{2}$.

$\mathcal{F}$ is a σ-algebra of events, i.e. $\mathcal{F} \subseteq \mathcal{P}(\Omega)$ and hence, $|\mathcal{F}| \leqslant 2^{|\Omega|} = 2^{\binom{N}{2}}$.

$\mathbb{P}$ is a measure on $(\Omega, \mathcal{F})$ called the probability measure. $\mathbb{P}(\Omega) = 1$.

In Section 2.1, we showed how to calculate node degrees based on a time window mechanism and we made the assumption that this is the only information available to nodes. We now place ourselves on the timeline, at the end of a window W. The current degree distribution reflects the contacts in all past windows and, if the network does not change erratically, we assume that it also provides a good estimate for the contacts in window $W + 1$. We will use this to calculate the probability distribution of C_t.

A degree distribution matches multiple graph instances. The probability of an edge existing in the network graph can be calculated as the number of graph instances containing that edge over the total number of graph instances matching our distribution. However, this is a difficult combinatorics problem. An easier way to calculate this probability is to use a graph construction model that generates fixed degree sequence graphs: the configuration model [2]. For every time window of size W, the configuration model guarantees the given degree distribution. In this model, each node is assigned a number of *stubs* or half edges, equal to its degree. The stubs are then paired with each other. The probability of an edge between nodes i and j is calculated as follows. For nodes i and j, and for $1 \leqslant s \leqslant d_i$ and $1 \leqslant t \leqslant d_j$, we define $I_{st,ij}$ to be the indicator of the event: "stub s is paired to the stub t", where the stubs are numbered arbitrarily. If $I_{st,ij} = 1$ for some st, then there is an edge between vertices i and j. It follows that the probability of an edge linking i and j is

$$p_{ij} = \sum_{\substack{1 \leqslant s \leqslant d_i \\ 1 \leqslant t \leqslant d_j}} Pr[I_{st,ij} = 1] = d_i d_j Pr[I_{11,ij} = 1], \tag{1}$$

since the probability of producing an edge between i and j by pairing the stubs s and t does not depend on s and t. Now, $Pr[I_{11,ij} = 1]$ is the probability that stubs 1 of i and 1 of j are paired to each other, which is equal to $\left(\sum_{i=1}^{N} d_i - 1\right)^{-1}$. Therefore, the probability of an edge between i and j is

$$p_{ij} = \frac{d_i d_j}{\sum\limits_{1 \leqslant i \leqslant N} d_i - 1} \tag{2}$$

and the probability of no edge between i and j is $1 - p_{ij}$.

Hence, the probability for an edge linking i and j to appear in the graph of the next time window is p_{ij} in equation 2. However, this is **not** the probability of the next contact being between i and j. Based on p_{ij}, and assuming uniform sampling of the edges generated by the configuration model (i.e., each node pair has the same contact frequency), this probability is shown in equation 3.

$$\mathbb{P}[(i,j)] = \frac{d_i d_j}{\sum\limits_{i=1}^{N} d_i - 1} \cdot \frac{1}{\frac{1}{2}\sum\limits_{i=1}^{N} d_i} = \frac{d_i d_j}{\frac{1}{2}\left(\sum\limits_{i=1}^{N} d_i\right)^2 - \frac{1}{2}\sum\limits_{i=1}^{N} d_i} = \frac{d_i d_j}{\sum\limits_{1 \leqslant i < j \leqslant N} d_i d_j + \sum\limits_{i=1}^{N} \binom{d_i}{2}} \tag{3}$$

The term $\sum_{i=1}^{N} \binom{d_i}{2}$ in the denominator of equation 3 reflects the probability of self-loops ("choose 2 stubs out of d_i" are the self-loops of node i) in the graph generated by the configuration model, which translates into contacts where only one node is involved. To avoid this, we consider the *erased configuration model* [24]. Starting from the multigraph obtained through the configuration model, we merge all multiple edges into a single edge and erase all self-loops. It was shown in [24] that, provided that the maximum degree of the graph is $o(\sqrt{N})$, the *configuration model* and the *erased configuration model* are asymptotically equivalent, in probability. Since we justifiably made this assumption in Section 2.1, we can safely approximate the probability of the next contact being between i and j as

$$\mathbb{P}[(i,j)] = \frac{d_i d_j}{\sum\limits_{1 \leqslant x < y \leqslant N} d_x d_y} \tag{4}$$

This concludes our estimation of the probability distribution of C_t, as we can now calculate the probability of any type of contact using the degree oracle.

3.2 Group Communication as a CCP

To analyse the problem in Definition 1 and its particular cases, we will use a notorious probabilistic method: the Coupon Collector Problem (CCP). This method was formally introduced in [11] and its setting is the following.

Lemma 1 (Coupon Collector). *Consider an unlimited supply of coupons of n distinct types. At each trial, we collect a coupon uniformly at random and independently of previous trials. Then, it takes on average nH_n trials, to collect at least one of each of the n coupon types (H_n is the Harmonic Number). In addition, the number of trials until the full collection is obtained is sharply concentrated around its expectation.*

In our problem, each contact is a trial, $\mathcal{L}$ - the set of carriers, is the collector and the nodes in $\mathcal{D}$ are the desired coupons. A node $d \in \mathcal{D}$ is collected when the first contact between d and any of the nodes in $\mathcal{L}$ has occurred. Unlike the coupon collector, each node has a different, known probability of being collected in our case. In [12], Flajolet et al. give formulas for the expected number of trials until both partial collections and a full collection, in the heterogeneous case.

Another difference between our setting and the CCP is the *null coupon*. The CCP has n types of coupons and $\sum_{k=1}^{n} p_n = 1$. We have D distinct nodes to be collected, plus a null event representing contacts in which none or both of the two nodes involved, carries the message m. In other words, there exist $n = D + 1$ types of coupons (D destination nodes and no message exchange) and successful message dissemination means collecting all but the null coupon, that is D coupons. Let p_0 be the probability of the null event: $p_0 + \sum_{k=1}^{D} p_k = 1$.

Lemma 2. *For each of the definitions in Section 2.2, the probabilities* p_i, $0 \leqslant i \leqslant D$ *are fully defined using the probability distribution from Section 3.1:*

$$p_k = \frac{d_{n_k} \sum_{j=1}^{L} d_{c_j}}{\sum_{1 \leqslant x < y \leqslant N} d_x d_y} \quad and \quad p_0 = \frac{\sum_{1 \leqslant i < j \leqslant L} d_{c_i} d_{c_j} + \sum_{1 \leqslant i < j \leqslant N-L} d_{n_i} d_{n_j} + \sum_{\substack{1 \leqslant i \leqslant L \\ 1 \leqslant j \leqslant N-L-D}} d_{c_i} d_{n_{D+j}}}{\sum_{1 \leqslant x < y \leqslant N} d_x d_y},$$

where c_1 to c_L are the L collector nodes, n_1 to n_D are the D^4 coupon nodes and n_{D+1} to n_{N-L} are the rest of the nodes.

Proof. The probability for the collector $\mathcal{L}$ to collect a fixed coupon type n_k, $1 \leqslant k \leqslant D$ is the probability that n_k meets any of the collector nodes, i.e.

$$p_k = \mathbb{P}\left[\bigcup_{1 \leqslant j \leqslant L} (n_k, c_j)\right] \overset{\text{mut. ex.}}{=} \sum_{1 \leqslant j \leqslant L} \mathbb{P}\left[(n_k, c_j)\right] \overset{\text{eq. 4}}{=} \frac{d_{n_k} \sum_{1 \leqslant j \leqslant L} d_{c_j}}{\sum_{1 \leqslant x < y \leqslant N} d_x d_y}. \tag{5}$$

Equation 5 defines all p_k $1 \leqslant k \leqslant D$. The null coupon probability is defined as follows (1st term: collectors meeting each other, 2nd term: non-collectors meeting each other, 3rd term: collectors meeting non-destinations):

$$p_0 = \mathbb{P}\left[\bigcup_{1 \leqslant i < j \leqslant L} (c_i, c_j) \cup \bigcup_{1 \leqslant i < j \leqslant N-L} (n_i, n_j) \cup \bigcup_{\substack{1 \leqslant i \leqslant L \\ 1 \leqslant j \leqslant N-L-D}} (c_i, n_{D+j})\right] \tag{6}$$

$$\overset{\text{mut. ex.}}{=} \sum_{1 \leqslant i < j \leqslant L} \mathbb{P}\left[(c_i, c_j)\right] + \sum_{1 \leqslant i < j \leqslant N-L} \mathbb{P}\left[(n_i, n_j)\right] + \sum_{\substack{1 \leqslant i \leqslant L \\ 1 \leqslant j \leqslant N-L-D}} \mathbb{P}\left[(c_i, n_{D+j})\right]$$

$$\overset{\text{eq. 4}}{=} \frac{\sum_{1 \leqslant i < j \leqslant L} d_{c_i} d_{c_j} + \sum_{1 \leqslant i < j \leqslant N-L} d_{n_i} d_{n_j} + \sum_{\substack{1 \leqslant i \leqslant L \\ 1 \leqslant j \leqslant N-L-D}} d_{c_i} d_{n_{D+j}}}{\sum_{1 \leqslant x < y \leqslant N} d_x d_y}.$$

Thus, we have a fully defined coupon collector probability vector. The vector is a function of the chosen set of *permanent message carriers*, i.e., the chosen collector $\mathcal{L}$ and it does not include the null coupon: $p(\mathcal{L}) = (p_1(\mathcal{L}), \dots, p_D(\mathcal{L}))$. ∎

---

4 We consider the case where $\mathcal{L} \cap \mathcal{D} = \varnothing$. Otherwise, some of the carriers (or all, if $D \leqslant L$) could be trivially assigned among the destinations.

We will now define an order relation for probability vectors of this type.

Lemma 3. *The binary relation $\leqslant$ defined on $X = \{p(\mathcal{L}) \mid \mathcal{L} \in \mathcal{P}(\mathcal{N}) \text{ and } |\mathcal{L}| = L\}$*

$$p(\mathcal{L}_a) \leqslant p(\mathcal{L}_b) \quad \Leftrightarrow \quad \min(p(\mathcal{L}_a)) \leqslant \min(p(\mathcal{L}_b)). \tag{7}$$

is a total order relation.

Proof. The minimum of a vector is a real number, hence our relation is equivalent to the total order relation $\leqslant$ on $\mathbb{R}$. Thus it is a total order relation. ∎

Using the total order relation, we frame a lemma on the monotonicity of $p(\mathcal{L})$.

Lemma 4. *The coupon collector probability vector, $p(\mathcal{L})$, is monotonous:*
    *Broadcast: $p(\mathcal{L})$ increases with $\sum_{c \in \mathcal{L}} d_c$, **with skewed degree distributions**.*
    *Multi-, Anycast: $p(\mathcal{L})$ **always** increases with $\sum_{c \in \mathcal{L}} d_c$.*

The proof and broadcast conditions for monotonicity are in the Appendix A.
    In the next section, we analyze the variable T_s. This represents the number of contacts until the D distinct coupons have been collected for broadcast and multicast and the number of contacts until one coupon has been collected for anycast.

3.3 Expected *-Cast Time

Using the probability distributions, we express the expected number of contacts until successful message dissemination. As before, the vector of collection probabilities is: $p(\mathcal{L}) = (p_1(\mathcal{L}), \ldots, p_D(\mathcal{L}))$, where $p_0 + \sum_{i=1}^{D} p_i = 1$.

Broadcast (Def. 2) and Multicast (Def. 3). For broadcast and multicast, the expected number of contacts until successful message dissemination is

$$\mathbb{E}[T_s] = \sum_{k=1}^{D} (-1)^{k+1} \sum_{1 \leqslant x_1 < \ldots < x_k \leqslant D} \frac{1}{p_{x_1} + \cdots + p_{x_k}} \overset{[7]}{=} \int_0^1 \left(1 - \prod_{i=1}^{D} (1 - t^{p_i})\right) \frac{dt}{t}. \tag{8}$$

by the inclusion-exclusion principle. Note that this formula holds regardless of whether $\sum_{k=1}^{D} p_k = 1$ holds (here, it does not). This function has been studied extensively in the mathematics literature and not only [21, 6, 1, 12]. In this study, we want to prove that it is positive and decreasing in p. As shown in Lemma 4, maximizing p amounts to maximizing the degrees of the carrier nodes. Therefore, we will also prove that choosing nodes of maximum degree as carriers is the optimum solution in terms of delivery delay.

Theorem 1. *With skewed degree distributions, for each $N \in \mathbb{N}$ and for each $L \in \mathbb{N}$, $L < N$, the function $f_D(p) = \mathbb{E}[T_s]$ is positive and decreasing in p, using the order relation defined in Lemma 3.*

Proof. By a simple change of variable ($t = e^{-u}$) in the integral in equation 8, $f_D(p)$ can also be written as

$$f_D(p) = \mathbb{E}[T_s] = \int_0^{\infty} \left(1 - \prod_{i=1}^{D} (1 - e^{-t p_i})\right) dt. \tag{9}$$

In [7], Borwein et al. showed that the key to proving an entire series of interesting properties of this function (including monotonicity) lies in a convenient way of writing it, derived from equation 9. Let X_i, $(1 \leqslant i \leqslant D)$ be independent positive exponentially distributed random variables with parameter $\lambda = 1$. Then,

$$f_D(p) = \mathbb{E}[T_s] = \int_0^\infty \left(1 - \prod_{i=1}^D \left(1 - e^{-tp_i}\right)\right) dt = \mathbb{E}\left(max\left(\frac{X_1}{p_1}, \ldots, \frac{X_D}{p_D}\right)\right). \qquad (10)$$

Then, from the representation in equation 10, it is straightforward that $f_D(p)$ is positive and decreasing with the increase of all p_1 to p_D.

In the case of *multicast* and *anycast*, all p_1 to p_D increase with the increase of $\sum_{c \in \mathcal{L}} d_c$ and the proof is finished.

For the *broadcast* case, all p_1 to p_D except one, will increase with the increase of $\sum_{c \in \mathcal{L}} d_c$ (see proof of Lemma 3). With a skewed degree distribution, the network has numerous nodes of minimum degree. This means, a considerable number of the probabilities p_1 to p_D will be equal and of minimum value. These minimum probabilities will have an overwhelming mass in $f_D(p)$. Together with all the other increasing probabilities, they will almost surely, largely outbalance the unique decreasing value of the vector $p(\mathcal{L})$.

Hence, $f_D(p)$ will always decrease with the increase of vector p, where the order of vectors $p(\mathcal{L})$ is defined as in equation 7 of Lemma 3. ■

Anycast (Def. 4). As far as anycast from Definition 4 is concerned, the problem is much simpler. Indeed, for the anycast case, one only need to deliver the message to one node in the destination set $\mathcal{D}$. Hence, the showing time, T_s is a geometric random variable with success probability

$$p_{any} = \mathbb{P}\left[\bigcup_{\substack{1 \leqslant i \leqslant D \\ 1 \leqslant j \leqslant L}} (n_i, c_j)\right] \overset{mut.\ ex.}{=} \sum_{\substack{1 \leqslant i \leqslant D \\ 1 \leqslant j \leqslant L}} \mathbb{P}[(n_i, c_j)] \overset{eq.\ 4}{=} \frac{\sum_{i=1}^D d_{n_i} \sum_{j=1}^L d_{c_j}}{\sum_{1 \leqslant x < y \leqslant N} d_x d_y}. \qquad (11)$$

Therefore, the expected value of T_s is

$$\mathbb{E}[T_s] = \frac{1}{p_{any}} = \frac{\sum_{1 \leqslant x < y \leqslant N} d_x d_y}{\sum_{i=1}^D d_{n_i} \sum_{j=1}^L d_{c_j}}, \qquad (12)$$

and it is obviously also decreasing with the increase of vector p, where the order of vectors $p(\mathcal{L})$ is defined as in equation 7. Finally, Theorem 1 and equation 12 allow us to conclude that choosing the nodes of maximum degree as permanent message carriers does indeed minimize the showing time portion of the delivery time in all message dissemination schemes considered.

4 Experimental Evaluation

We provide here a brief experimental verification of our main finding in Section 3.3: having maximum degree nodes as carriers minimizes the showing time.

4.1 Contact Generators

We use three contact generators: two real mobility traces and one synthetic. The synthetic contacts are based on a scale-free graph model. The real mobility traces originate from two data collection experiments conducted by universities.

Scale Free Contacts: The contacts whereof the aggregation results in a scale free (SF) social graph are generated as follows. First, each node is assigned a popularity according to a power law (exponent 3). Then, the two nodes participating in a contact are randomly and independently chosen according to their respective popularities. We use a scale free model with 500 nodes and 50 000 contacts.

ETH Contacts: The first trace comes from an experiment of ETH Zürich [18]. 20 students and staff working on the same floor of an ETH building carried 802.11-enabled devices for 5 days. Every 0.5 s, each device sent a beacon message, the reception of which was logged by all devices in 802.11 radio proximity. This trace contains more than 23 000 reported contacts and is unique in terms of time granularity and reliability. Although the ETH trace measurement period spans a relatively short time, there are on average more than 1 000 contacts per device. This is comparable in number of contacts to similar longer traces.

MIT Contacts: The second trace comes from the Reality Mining [10] project. 97 students and employees of MIT were equipped with mobile phones scanning every 5 minutes for Bluetooth devices in proximity during 9 months. This trace is unique in terms of number of devices and duration. Nevertheless, with a time granularity of 5 minutes, many short contacts were presumably not logged. For our simulations, we cut the trace at both ends and used 100 000 contacts reported between September 2004 and March 2005. Note that this time period contains holidays and semester breaks and thus still captures varying user behavior.

For both real mobility traces we ignored logged timing information and just ordered the reported contacts according to their start times (i.e., slotted contacts). We obtain the social graph from the contacts using the method discussed and the results (optimal parameters) obtained in [14].

4.2 Simulation Results

To confirm our analysis, we evaluated the performance of constrained broadcast and multicast as described in Definitions 2 and 3, for Uniform Randomly chosen carriers (UR) versus message carriers with Highest Degree (HD) in the network.

For each contact generator, the simulation has a warmup period, to allow the collection of information and a cool down period, to allow the messages created last to be in the simulation for one TTL, as well. The TTLs were found empirically, with the aim of a reasonable coverage. Messages are generated randomly with probability 0.05 at each contact, during the period between the warmup and the cool down times. For multicast, we consider groups of size 25% of all nodes, chosen uniformly at random.

The two metrics considered are the **coverage**: the percentage of nodes in $\mathcal{D}$ receiving the message within TTL steps, and the **number of delivered messages** over time. This last metric clearly captures the average delay of the two strategies, as well.

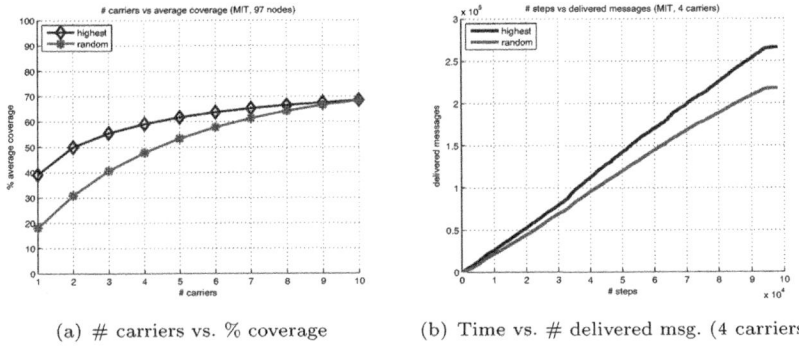

(a) # carriers vs. % coverage   (b) Time vs. # delivered msg. (4 carriers)

Fig. 1. Broadcast Results (MIT trace)

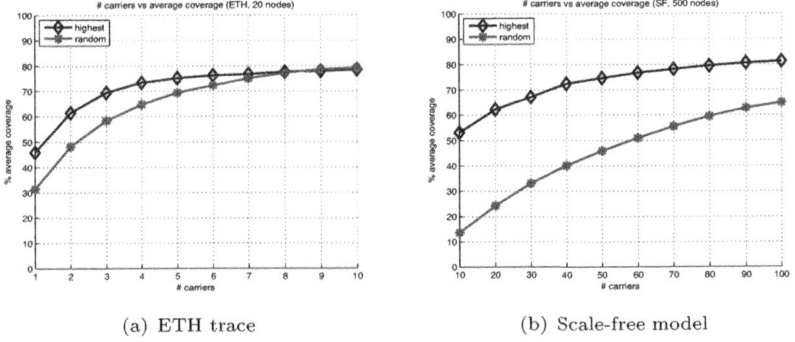

(a) ETH trace   (b) Scale-free model

Fig. 2. Multicast Results: # carriers vs. % coverage

Figure 1 shows the two metrics for broadcast using the MIT trace. Plots for the ETH trace and the scale-free model are qualitatively similar and were left out due to space limitations. Figure 1(a) shows that the coverage is indeed better with the HD strategy. However, as the number of carriers approaches the total number of nodes in the network, the two curves converge, since the two sets of carriers (HD based and UR chosen) overlap more and more.

Figure 1(b) shows the number of messages delivered during the trace, for one point in Figure 1(a). It confirms that the HD scheme does indeed take less time than the UR scheme to deliver the same amount of messages. Figures 2(a) and 2(b) show the coverage for multicast on the ETH trace and respectively, the scale-free model. Again, HD has better coverage and delivery time in both cases.

As a final note, we observed that, using the two traces with a higher numbers of carriers, often the UR scheme slightly outperforms the HD scheme. We believe it is an effect of one social relation, our current analysis does not account for: node communities. Indeed, if all high degree nodes are part of the same community, the remaining communities will not have any member who is a carrier and will thus have a low chance of receiving the message. The UR scheme, however, has better odds of sampling carriers from smaller communities, hence the better performance with sufficient carriers. Accounting for node communities is the next step in our work.

5 Conclusion

Our goal in this paper has been to take a first look into a neglected area of DTN routing: *-cast. More precisely, we have undertaken a study of broad-, multi- and anycast in DTNs using social information. We identified a class of relevant optimization problems for *-cast, where the amount of resources (message copies) is constrained, and the optimization goal is to allocate resources to relays so as to minimize the delay until all nodes in the target *-cast group receive the message.

To solve these problems, we used social network analysis to render complex mobility patterns into a graph, transforming the problem into choosing an optimal set of vertices. Our first contribution is a probabilistic model that maps this graph to future contact probabilities between nodes. Moreover, in a setting where the degree distribution of this graph is skewed and known a priori, we prove, using a coupon collector analogy, that when the highest degree nodes are message relays the expected *-cast delivery time is optimum. We corroborate our findings using an evaluation of this policy on both synthetic and real mobility traces.

This is merely a first step into an area of many interesting, open issues. With the insight gained from this study, our next aim is to prove optimal policies for the problem where nodes are met online and degrees of nodes not yet met are unknown. Moreover, in practice, social graphs are expected to have high clustering coefficients and community structure. This makes the choice of highest degree nodes as carriers not necessarily optimal, due to the potential similarities among the neighborsets of high degree nodes. We are currently studying optimal policies accounting for both degrees and communities to optimally allocate resources.

References

1. Adler, I., Oren, S., et al.: The coupon-collector's problem revisited. J. Appl. Probab. 40(2), 513–518 (2003)
2. Aiello, W., Chung, F., et al.: A random graph model for massive graphs. In: STOC 2000, pp. 171–180. ACM, New York (2000)
3. Albert, R., Barabási, A.L.: Statistical mechanics of complex networks. Rev. Mod. Phys. 74(1), 47–97 (2002)
4. Altman, E., Başar, T., et al.: Optimal monotone forwarding policies in delay tolerant mobile ad-hoc networks. In: ValueTools 2008, pp. 1–13. ICST, Brussels (2008)
5. Boldrini, C., Conti, M., et al.: ContentPlace: social-aware data dissemination in opportunistic networks. In: MSWiM 2008, pp. 203–210. ACM Press, New York (2008)
6. Boneh, S., Papanicolaou, V.: General asymptotic estimates for the coupon collector problem. J. Comput. Appl. Math. 67(2), 277–289 (1996)
7. Borwein, J., Affleck, I., et al.: Convex? Convex! Convex II. SIAM: Problems and Solutions (2000)
8. Daly, E., Haahr, M.: Social network analysis for routing in disconnected delay-tolerant MANETs. In: MobiHoc 2007, pp. 32–40. ACM Press, New York (2007)
9. Dubois-Ferriere, H., Grossglauser, M., et al.: Age matters: efficient route discovery in mobile ad hoc networks using encounter ages. In: MobiHoc 2003, pp. 257–266. ACM Press, New York (2003)
10. Eagle, N., Pentland, A.: Reality mining: sensing complex social systems. Personal Ubiquitous Comput. 10(4), 255–268 (2006)

11. Feller, W.: An introduction to probability theory and its applications, 2nd edn., vol. 2. John Wiley & Sons, Chichester (1957)
12. Flajolet, P., Gardy, D., et al.: Birthday paradox, coupon collectors, caching algorithms and self-organizing search. Disc. Appl. Math. 39(3), 207–229 (1992)
13. Gao, W., Li, Q., et al.: Multicasting in delay tolerant networks: a social network perspective. In: MobiHoc 2009, pp. 299–308. ACM, New York (2009)
14. Hossmann, T., Legendre, F., et al.: From Contacts to Graphs: Pitfalls in Using Complex Network Analysis for DTN Routing. In: INFOCOM NetSciCom 2009, pp. 1–6 (2009)
15. Hui, P., Chaintreau, A., et al.: Pocket switched networks and human mobility in conference environments. In: SIGCOMM WDTN 2005, pp. 244–251. ACM, New York (2005)
16. Hui, P., Crowcroft, J., et al.: Bubble rap: social-based forwarding in delay tolerant networks. In: MobiHoc 2008, pp. 241–250. ACM Press, New York (2008)
17. Leguay, J., Friedman, T., et al.: Evaluating Mobility Pattern Space Routing for DTNs. In: INFOCOM 2006, pp. 1–10 (2006)
18. Lenders, V., Wagner, J., et al.: Measurements from an 802.11b Mobile Ad Hoc Network. In:WOWMOM 2006, pp. 519–524. IEEE Computer Society, Washington (2006)
19. Lindgren, A., Doria, A., et al.: Probabilistic routing in intermittently connected networks. SIGMOBILE Mob. Comput. Commun. Rev. 7(3), 19–20 (2003)
20. Mihail, M., Saberi, A., et al.: Random walks with lookahead on power law random graphs. Internet Math. 3(2), 147–152 (2006)
21. Nath, H.: Waiting Time in the Coupon-Collector's Problem. Austr. & New Zeal. J. of Stat. 15(2), 132–135 (1973)
22. Newman, M.E.J.: The structure and function of complex networks. SIAM Review 45, 167 (2003)
23. Spyropoulos, T., Psounis, K., et al.: Spray and wait: an efficient routing scheme for intermittently connected mobile networks. In: SIGCOMM WDTN 2005, pp. 252–259. ACM Press, New York (2005)
24. van der Hofstad, R.: Random Graphs and Complex Networks (2009)
25. Zhang, Z.: Routing in intermittently connected mobile ad hoc networks and delay tolerant networks: overview and challenges. IEEE Communications Surveys & Tutorials 8(1), 24–37 (2007)
26. Zhao, W., Ammar, M., et al.: Multicasting in delay tolerant networks: semantic models and routing algorithms. In: SIGCOMM WDTN 2005, pp. 268–275. ACM, New York (2005)

A Proof of Lemma 4

Proof. We want to prove that, as we increase $\sum_{c \in \mathcal{L}} d_c$, $p(\mathcal{L})$ will also increase, which in light of Lemma 3, means that $\min(p(\mathcal{L}))$ will increase.

Suppose we replace one collector node, c_i with one non-collector node n_j. Denote the new collector set $\mathcal{L}'$, n_j is relabeled c_i' and c_i is relabeled n_j'. Assume $d_{c_i} < d_{n_j}$. It follows that

$$\sum_{c \in \mathcal{L}} d_c < \sum_{c \in \mathcal{L}'} d_c, \tag{13}$$

We must prove that $\min(p(\mathcal{L})) < \min(p(\mathcal{L}'))$.

Multi-, Anycast: We start with the easier case of multicast and anycast, with $\mathcal{L} \cap \mathcal{D} = \emptyset$. Otherwise, some of the carriers (or all, if $D \leqslant L$) could be trivially assigned among the destinations. Under this condition, $c_i \notin \mathcal{D}$. Therefore, n_j'

will not be added to the destination set and there will be no extra p'_j. Whether n_j is a destination or not is irrelevant, as it becomes a collector node and will receive a copy of the message for storage. By equations 13 and 5, **p_1 to p_D will all increase** and thus, so will their minimum $\min(p(\mathcal{L}'))$. That is $\min(p(\mathcal{L})) < \min(p(\mathcal{L}'))$ and the proof for multicast and anycast in Lemma 4 is finished.

Broadcast: In broadcast, $\mathcal{D} = \mathcal{N}$, therefore, all nodes are also destinations, consequently $c_i \in \mathcal{D}$. This makes the evolution of p_1 to p_D slightly more ambiguous than previously. Whereas above, the destination set is not affected by the replacement, here the destination set, $\mathcal{N} \setminus \mathcal{L}$, is inevitably altered.

In particular, node c_i will now be part of the destination set, whereas it was not, before the replacement. Removing c_i from the collector set means, c_i will no longer get a copy of the message for storage and it will have to be counted in the probabilities p_1 to p_D, i.e., there will be a new probability, p'_j. **The original probabilities will all increase** by equations 5 and 13. The fate of the new minimum, $p'_{\min} = \min(p(\mathcal{L}'))$, depends on $p'_j = \dfrac{d_{n'_j} \sum_{c \in \mathcal{L}'} d_c}{\sum\limits_{1 \leqslant x < y \leqslant N} d_x d_y}$. Denote by $\min(d_{\mathcal{N} \setminus \mathcal{L}})$ and respectively $\min(d_{\mathcal{N} \setminus \mathcal{L}'})$, the minimum degree among the nodes in the destination set. Then, there are two possibilities:

a. $d_{n'_j} \geqslant \min(d_{\mathcal{N} \setminus \mathcal{L}})$. The return of n'_j to the destination set does not change the minimum $\min(p(\mathcal{L}))$. It will have increased, but it will still correspond to the same node as before the replacement. Therefore, $p(\mathcal{L}) \leqslant p(\mathcal{L}')$.

b. $d_{n'_j} < \min(d_{\mathcal{N} \setminus \mathcal{L}})$. The return of n'_j to the destination set could change the minimum. This can be determined by checking the sign of $p_{\min} - p'_{\min}$, the difference between the minimum before and respectively after the replacement. Denote $C = \sum_{c \in \mathcal{L} \setminus \{c_i\}} d_c = \sum_{c \in \mathcal{L}' \setminus \{c'_i\}} d_c$ and $X = \left(\sum_{1 \leqslant x < y \leqslant N} d_x d_y \right)^{-1}$. According to equation 5, $p_{\min} = \min(d_{\mathcal{N} \setminus \mathcal{L}}) \cdot \left(C + d_{n'_j} \right) \cdot X$ and $p'_{\min} = d_{n'_j} \cdot \left(C + d_{c'_i} \right) \cdot X$. Then,

$$p_{\min} - p'_{\min} = \left[C(\min(d_{\mathcal{N} \setminus \mathcal{L}}) - d_{n'_j}) + d_{n'_j}(\min(d_{\mathcal{N} \setminus \mathcal{L}}) - d_{c'_i}) \right] \cdot X. \qquad (14)$$

From the assumption of item b, the first term of the sum in equation 14 is positive. Moreover, as $\min(d_{\mathcal{N} \setminus \mathcal{L}})$ was the minimum before the replacement, clearly $\min(d_{\mathcal{N} \setminus \mathcal{L}}) \leqslant d_{n_j} = d_{c'_i}$ and thus, the second term of the sum in equation 14 is negative. This means that the monotonicity of $p(\mathcal{L})$ might not hold for a certain choice of parameters, when the positive term in equation 14 outweighs the negative term. However, as the discussion below will show, in practice, the monotonicity is always respected.

Discussion. To sum up the above proof, probability vectors $p(\mathcal{L})$ increase with $\sum_{c \in \mathcal{L}} d_c$, as per the order relation in Lemma 3, with one exception: item b of the broadcast case. Under an arbitrary degree distribution, this implies it might be optimal to have very low degree nodes as message carriers (i.e., they receive a message copy to store). This is merely an artifact of the fact that, currently, we analyze the showing time, T_s alone. It will be reconciled when we consider both the finding time, T_f and the showing time, T_s together. In practice, degree distributions are, more often than not, skewed. A skewed degree distribution means there are numerous minimum degree nodes. As is evident from equation 14, giving messages copies to small degree nodes worsens the probability vector. ∎

Applying Branching Processes to Delay-Tolerant Networks

Dieter Fiems[1] and Eitan Altman[2]

[1] SMACS Research Group, Department TELIN, Ghent University, Belgium
[2] INRIA, BP93, 06902 Sophia Antipolis, France

Abstract. Mobility models that have been used in the past to study delay tolerant networks (DTNs) have been either too complex to allow for deriving analytical expressions for performance measures, or have been too simplistic. In this paper we identify several classes of DTNs where the dynamics of the number of nodes that have a copy of some packet can be modeled as branching process with migration. Using recent results on such processes in a random environment, we obtain explicit formulae for the first two moments of the number of copies of a file that is propagated in the DTN, for quite general mobility models. Numerical examples illustrate our approach.

1 Introduction

Delay tolerant networks (DTNs) embrace the concept of occasionally-connected networks [8,11], such as sensor networks, wireless networks with alternating connectivity, etc. In this paper, we address packet forwarding in DTNs where connectivity is low and nodes relay packets of other nodes. We focus on two-hop routing schemes [16] in which a relay node that receives a packet from the source does not relay it further to other intermediate nodes. (Such a restriction may be needed in the context of resource limitations or for security reasons.) We show that various dynamics of packet forwarding in DTNs can be described by multi-type branching processes with immigration operating in a random environment. We then use novel tools from branching processes with immigration in order to derive the two first moments of the number of nodes with a copy of the file.

Related work. Before proceeding to the main results, we present a brief overview of the scientific context of the branching processes methodology and to their applications in networking. The first results on branching processes are often attributed to Galton and Watson and date back to the 19th century. At that time, there was a severe concern among aristocratic families that the surnames were becoming extinct. The disappearance of a name of a family was considered as the death of the family and it was thought that the extinct families were replaced by families from lower social layers [2]. F. Galton posed the question of computing the extinction probability of the names in the Educational Times of 1873 [14]. More precisely, assume that each man in generation n has some

E. Altman et al. (Eds.): Bionetics 2009, LNICST 39, pp. 117–125, 2010.

random number of sons in generation $n + 1$, according to a fixed probability distribution that does not vary from individual to individual. What is then the probability that a family dies out? The Reverend Henry William Watson replied with a solution [27]. Together, they then wrote an 1874 paper entitled "On the probability of extinction of families" [15]. Galton and Watson appear to have derived their process independently of the much earlier work by the French statistician I. J. Bienaymé [6] (1845), which was unknown till it was rediscovered in 1962 by Heyde and Senneta, see e.g. [19].

Branching processes with a random environment have been well studied, both with and without immigration, see [5]. For example, conditions are presented for the extinction when the random environment is stationary ergodic. The stability, strong law of large numbers and central limit theorems for multi-type branching processes with immigration in a random environment have been studied in [20,26]. These processes find applications in very diverse fields, including biological systems and queueing theory. For example, McNamara et. Al [23] consider an asexual species with non-overlapping generations. Individuals born in some year, reach maturity and reproduce one year later and then die. The number of individuals of the different genotypes in the consecutive years constitute a multi-type branching process. Prime examples in queueing theory where branching processes with immigration play a major role, include infinite server queues [10], processor sharing queues [17,24], as well as various polling systems [4,25]. The infinite server queue with random environment has been studied recently in [9,12]. These authors assume a independent exponentially distributed interarrival and service times. The theoretical framework applied here allows for explicit expressions for the first and second moments in the more general setting of general stationary ergodic processes describing the contact processes between pairs of nodes and general independent bounded service time, with a Markovian random environment. It builds on the Theory we developed in [13] and in references therein that allows to compute explicitly the two first moments of the branching process for the case of general stationary ergodic immigration process.

2   Theoretical Framework

First, we briefly present the standard (basic) scalar branching process taking integer values. We then present several extensions, including the vector (multi-type) case. In particular, we introduce the framework of [13] that extends branching processes, and yet provides explicit expressions for the first two moments.

2.1   The Scalar Integer-Valued Case

The standard branching is defined as follows. Let X_n be the number of individuals in generation n. Starting with a fixed X_0, we define recursively

$$X_{n+1} = \sum_{i=1}^{X_n} \xi_n^{(i)} \tag{1}$$

where $\xi_n^{(i)}$ are independent and identically distributed random variables taking non-negative integer values. Define $A_n(m) := \sum_{i=1}^{m} \xi_n^{(i)}$ we can rewrite the above as

$$X_{n+1} = A_n(X_n). \qquad (2)$$

Branching processes with immigration are defined through the recursion

$$X_{n+1} = A_n(X_n) + B_n. \qquad (3)$$

From equations (1) and (2), A_n obviously possess a divisibility property; for any non-negative integers m, m_1 and m_2 such that $m_1 + m_2 = m$, and for any n,

$$A_n(m) = A_n^{(1)}(m_1) + A_n^{(2)}(m_2)$$

where for each n, $A_n^{(1)}$ and $A_n^{(2)}$ are independent random processes, both with the same distribution as A_n.

This divisibility property naturally leads to the definition of branching processes on a continuous state space. We take this property, together with the non-negativity of A_n as the basis to define the continuous state branching processes. Noting that these properties are satisfied by Lévy processes, we define a continuous state branching process as one satisfying (2) where A_n is a non-negative Lévy process. For references as well as for alternative (equivalent) definitions, see [1,7,21,22] and the references therein.

2.2   General Setting

Consider the sequence of random column vectors $X_n \in \mathbb{R}^M$, adhering to,

$$X_{n+1} = A_n(X_n, Y_n) + B_n(Y_n), \quad n \in \mathbb{Z}, \qquad (4)$$

The process Y_n and the vector valued processes A_n and B_n correspond to the environment process, the branching process and the immigration process, respectively. The random environment Y_n is a stationary ergodic Markov chain, taking values on a finite state-space $\Theta = \{1, 2, \ldots, N\}$; let $P = [p_{ij}]$ denote its transition matrix. The branching processes $A_n : \mathbb{R}^M \times \Theta \to \mathbb{R}^M$ are independent and identically distributed and further adhere to the following assumptions.

- For each $i \in \Theta$, $A_n(\cdot, i)$ has a divisibility property. Let $x = x^1 + x^2 + \ldots + x^k \in \mathbb{R}^M$, then $A_n(x, i)$ has the following representation,

$$A_n(x, i) = \sum_{l=1}^{k} \hat{A}_n^{(l)}(x^l, i), \qquad (5)$$

whereby $\hat{A}_n^{(l)}(\cdot, i), l = 1, \ldots, k$, are identically distributed, but not necessarily independent, with the same distribution as $A_n(\cdot, i)$. Branching processes are those in which $\hat{A}_n^{(l)}(\cdot, i), l = 1, \ldots, k$, are independent.

– For each $i \in \Theta$ and $x = [x_1, \ldots, x_M] \in \mathbb{R}^M$, the first and second order moments of $A_n(\cdot, i)$ can be expressed as follows,

$$\mathrm{E}[A_n(x,i)] = \mathcal{A}_i x, \quad \mathrm{E}[A_n(x,i)A'_n(x,i)] = F_i(xx') + \sum_{j=1}^{M} x_j \Gamma_{i,j}, \quad (6)$$

whereby $\mathcal{A}_i$ and $\Gamma_{i,j}$ are fixed $M \times M$ matrices and F_i is a linear operator that maps $M \times M$ non-negative definite matrices on $M \times M$ non-negative definite matrices and satisfies $F_i(0) = 0$.

Finally, the immigration process $B_n : \Theta \to \mathbb{R}^M$ is a stationary ergodic sequence of random functions. The first and second order moments are denoted by $b_i = \mathrm{E}[B_0(i)]$ and $\mathcal{B}_{ij}^{(n)} = \mathrm{E}[B_0(i)B_n(j)]$.

Before proceeding to the main theorems, some notation is introduced. Let $\hat{A}$ denote the block matrix whose ijth block entry is given by $\mathcal{A}_j p_{ji}$ $(i, j \in \Theta)$. Moreover, the following block vector and block matrix simplify notation,

$$\hat{b} = \sum_{i \in \Theta} \pi_i \begin{bmatrix} p_{i1}b_i \\ p_{i2}b_i \\ \vdots \\ p_{iN}b_i \end{bmatrix}, \quad \hat{\mathcal{B}}^{(n)} = \sum_{i \in \Theta} \pi_i \begin{bmatrix} \mathcal{B}_{i1}^{(n)}p_{i1} & \mathcal{B}_{i2}^{(n)}p_{i1} & \cdots & \mathcal{B}_{iN}^{(n)}p_{i1} \\ \mathcal{B}_{i1}^{(n)}p_{i2} & \mathcal{B}_{i2}^{(n)}p_{i2} & \cdots & \mathcal{B}_{iN}^{(n)}p_{i2} \\ \vdots & & \ddots & \\ \mathcal{B}_{i1}^{(n)}p_{iN} & \mathcal{B}_{i2}^{(n)}p_{iN} & \cdots & \mathcal{B}_{iN}^{(n)}p_{iN} \end{bmatrix}. \quad (7)$$

The existence of a stationary solution is now asserted by the following theorem. Theorem 2 then provides expressions for the first and second order moments of this solution. The proofs of these theorems can be found in [13].

Theorem 1. *Assume that (i) $b_i < \infty$ component-wise for all $i \in \Theta$; and (ii) that all the eigenvalues of the matrix $\hat{A}$ are within the open unit disk. Then, there exist a unique stationary solution X_n^*, for $n \in \mathbb{Z}$ such that $\lim_{n\to\infty} \|X_n - X_n^*\| = 0$, almost surely, for any initial value X_0.*

Theorem 2. *Assume that the conditions of Theorem 1 are satisfied. The conditional first moment vector is then given by,*

$$\mu = [\mathrm{E}[X_0^* 1\{Y_0 = i\}]]_{i \in \Theta} = (\mathcal{I} - \hat{A})^{-1}\hat{b}. \quad (8)$$

Under the additional assumption that the second order moments of $B_0(i)$ are finite, $i \in \Theta$, the elements Ω_i of the conditional second moment matrix of X_0^* are the unique solution of the system of equations,

$$\Omega_l = \mathrm{E}[X_0^*(X_0^*)' 1\{Y_0 = l\}]$$
$$= \sum_{k \in \Theta} \left(F_k(\Omega_k) + \sum_{j=1}^{M} \mu_k^{(j)} \Gamma_k^{(j)} + \mathcal{B}_{kk}^{(0)} \pi_k + \mathcal{A}_k \Lambda_k + \Lambda_k' \mathcal{A}_k' \right) p_{kl}, \quad (9)$$

$l \in \Theta$, *where Λ_k denotes the kth diagonal (block) element of $\sum_{j=0}^{\infty} \hat{A}^j \hat{\mathcal{B}}^{(j+1)}$ and with $\mu_k^{(j)}$ the jth element of $\mu_k = \mathrm{E}[X_0^* 1\{Y_0 = k\}]$.*

3   DTNs with Variable Number of Nodes

Consider a sparse content distribution network with mobile nodes. At each time slot, new mobile nodes may join or may leave this network. A fixed node spreads some content to other nodes of this network. The goal of the network is to offer access to that content to potential (mobile or fixed) clients that may request it. Whenever the source is within the transmission range of another node, it transmits a packet to that node. A two-hop routing scheme is adopted [16]. A relay node that receives a packet from the source does not relay it further to other intermediate nodes of the network. It only delivers it to a client whenever it encounters one. Time is discrete and at each time n, each node has a probability $p_\theta \geq p > 0$ to meet the source node; this probability also depends on the state $\theta \in \Theta$ of a modulating Markov chain. The chain allows us to model correlation between the channel conditions of different mobiles: it models global fluctuations in the channel conditions that affect the whole system simultaneously. For example, if it rains, then the probability that transmission from the source to a mobile j is successful will dicrese for all mobiles. A measure of the efficiency of the network in distributing the content is then the expected number of mobiles that have a copy of the content (packet) as well as its second moment.

Let W_n denote the number of nodes that have the packet at slot n and let Z_n denote the number of nodes that do not have the packet. We have the following recursion,

$$W_{n+1} = \sum_{j=1}^{W_n} \zeta_{n,1}^{(j)} + \sum_{j=1}^{Z_n} \zeta_{n,2}^{(j)} \nu_n^{(j)}, \ \ Z_{n+1} = \sum_{j=1}^{Z_n} \zeta_{n,2}^{(j)}(1 - \nu_n^{(j)}) + B_n.$$

Here $\zeta_{n,1}^{(j)}$ is the indicator that the jth node that has the packet leaves the system at slot n, $\zeta_{n,2}^{(j)}$ is the indicator that the jth node that does not have the packet leaves the system at slot n and $\nu_n^{(j)}$ is the indicator that the jth node that does not have the packet, receives the packet at slot n. Finally, B_n denotes the number of new nodes that arrive during slot n. Assuming stationary ergodic arrivals of nodes and independent geometrically distributed residence times with mean T, the theoretical framework applies.

Assuming a Markovian environment with two states, its transition probabilities are characterised by the fraction σ that the environment is in state 1 and by the mean time τ to alternate from state 1 to state 2 and back. Figure 1 depicts the mean number of nodes E[W] that have the packet and the mean number of nodes E[Z] that do not have the packet. The left pane plots these means vs. T for different values of τ. The mean number of nodes in the system is fixed to 50 by scaling the mean number of arrivals E[B] in a slot for increasing T. In state 2, a node receives the packet with probability $p_2 = 0.1$ whereas no transmission is possible in state 1 ($p_1 = 0$). Moreover, for all curves, $\sigma = 90\%$. It is readily observed that the mean residence time of a node has a considerable impact on E[W]. Obviously, if nodes remain longer, they carry the packet for a longer time which explains the increase in the mean number of nodes that carry the packet.

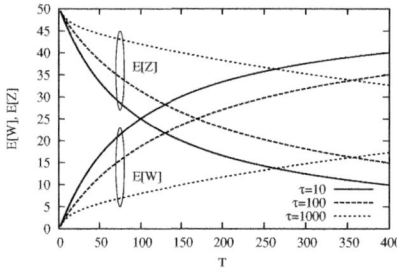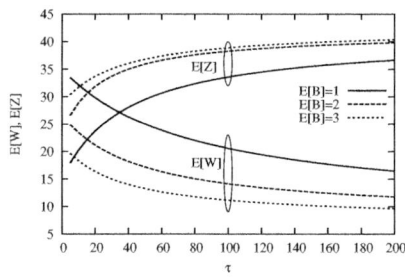

Fig. 1. Mean number of nodes that have the packet and of the number of nodes that do not have the packet vs. T for various values of τ (left) and vs. τ for various values of E[B] (right).

Further, increasing τ yields lower values of E[W]. This is confirmed by the right pane where E[Z] and E[W] are depicted vs. τ for various values of E[B] and the same parameter settings.

4 Mobility of the Source and the Nodes

We retain the model of the previous section but now replace the channel model by a mobility model. The source node moves according to a random walk through the spatial grid, depicted in Figure 2 (left). In each of the regions of the grid, new nodes arrive according to a stationary ergodic process which then travel through the grid until they leave. If a node is in the same region as the source, the node receives the packet with a fixed (possibly region-dependent) probability.

Let $X_n(k)$ denote the number of nodes in region k at time n with the packet and let $Z_n(k)$ denote the number of nodes without the packet. Further let X_n and Z_n denote the column vectors with elements $X_n(k)$ and $Z_n(k)$, respectively. Let Y_n denote the region where the source node resides at time n — the environment thus tracks the position of the source node — and let $B_n(k)$ denote the number of new nodes that arrive in region k at time n; B_n is a column vector with elements $B_n(k)$. We then have the following recursion,

$$X_{n+1} = \sum_{i=1}^{N} \sum_{j=1}^{X_n(i)} \zeta_{n,1}^{(i,j)} + \sum_{i=1}^{N} \sum_{j=1}^{Z_n(i)} \zeta_{n,2}^{(i,j)} \nu_n^{(i,j)}, \quad Z_{n+1} = \sum_{i=1}^{N} \sum_{j=1}^{Z_n(i)} \zeta_{n,2}^{(i,j)} (1 - \nu_n^{(i,j)}) + B_n.$$

Here $\zeta_{n,1}^{(i,j)}$ is a column vector of indicators; its kth element is the indicator that the jth node in region i that has the packet at time n moves to region k. The indicator vector $\zeta_{n,2}^{(i,j)}$ is defined likewise. Its kth element is the indicator that the jth node in region i that does not have the packet at time n moves to region k. Further, $\nu_n^{(i,j)}$ denotes the indicator that the jth node in region i that does not have the packet at time n, receives the packet. Notice that some of the packets may leave the grid as not all packets necessarily move to any of the regions.

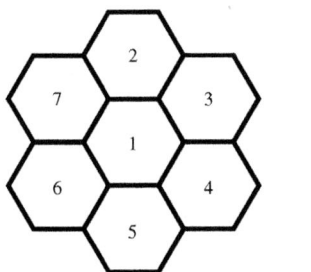 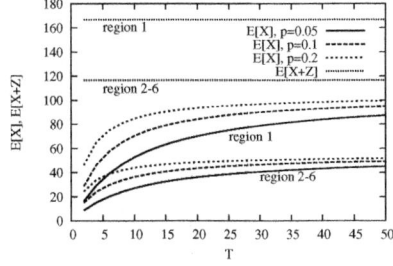

Fig. 2. Spatial grid of the nodes (left) and mean number of nodes with the packet and mean number of nodes vs. the mean residence time in a region (right)

Assuming geometrically distributed residence times (possibly region dependent) and random routing between the regions, the theoretical framework is applicable.

To limit the number of parameters involved, we assume that the mean residence times in the different regions are equal and nodes move to any of the neighbouring regions with probability $1/6$, thereby possibly leaving the grid (however, the source node never leaves the grid). A node that does not have the packet which is in the same region as the source node, receives the packet with probability p. The right pane of Figure 2 depicts the mean number of nodes with the packet in the different regions is vs. T for different values of the transmission probability p. The mean numbers of new arrivals in the different regions scale with the residence times of the nodes: $E\,B^{(i)} = 50/T$ for $i = 1, 2, \ldots, 7$ such that the total mean number of nodes in the different regions $E[X + Z]$ remains constant. First, notice that by symmetry, the characteristics of regions 2 to 7 are the same. Further, it is clear that longer residence times imply that more nodes receive the packet. Clearly, nodes do not only remain longer in a region but also longer in the grid. Hence, the probability that they receive the packet increases.

5   Conclusions

In this paper, it was shown that some of the dynamics of packet forwarding in DTNs can be described by Markov-modulated branching processes with immigration. The paper illustrates how explicit expressions for the two first moments of the state in DTNs can be obtained. This extends previously known time homogeneous models (without the ramdom environment feature) for which explicit expressions are known for relevant performance measures in DTNs [3,18].

Acknowledgement

This work was partly supported by the EuroNF European Network of Excellence. The work of the second author was supported partly by the BIONETS European project.

References

1. Adke, S.R., Gadag, V.G.: A new class of branching processes. In: Heyde, C.C. (ed.) Branching Processes: Proceedings of the First World Congress. Springer Lecture Notes, vol. 99, pp. 1–13 (1995)
2. Albertsen, K.: The Extinction of Families. International Statistical Review/Revue Internationale de Statistique 63(2), 234–239 (1995)
3. Altman, E.: Competition and cooperation between nodes in Delay Tolerant Networks with Two Hop Routing. In: Proceedings of Netcoop 2009, Eindhoven, The Netherlands (November 2009)
4. Altman, E., Fiems, D.: Expected waiting time in polling systems with correlated vacations. Queueing Systems 56(3-4), 241–253 (2007)
5. Athreya, K., Vidyashankar, A.: Branching Processes. In: Maddala, G., Rao, C., Vinod, H. (eds.) Handbook of Statistics, ch. 2, vol. 19. Elsevier Science B.V., Amsterdam (2001)
6. Bienaymé, I.J.: De la loi de la multiplication et de la durée des familles. Soc. Philomath., Paris Extraits Ser. 5, 37–39 (1845)
7. Bertoin, J.: Lévy Processes. Cambridge University Press, Cambridge (2002)
8. Cerf, V., et al.: Delay-tolerant network architecture. IETF RFC 4838
9. D'Auria, B.: $M/M/\infty$ queues in semi-markovian random environment. Queueing Systems 58(3), 221–237 (2008)
10. Eliazar, I.: On the discrete-time $G/GI/\infty$ queue. Technical report, Holon Institute of Technology (2006)
11. Fall, K.: A delay-tolerant network architecture for challenged internets. In: Proceedings of SIGCOMM 2003, pp. 27–34 (2003)
12. Falin, G.: The $M/M/\infty$ queue in a random environment. Queueing Systems 58(1), 65–76 (2008)
13. Fiems, D., Altman, E.: Markov-modulated stochastic recursive equations with applications to delay-tolerant networks. INRIA Research Report No. 6872 (2009); Submitted to Performance Evaluation
14. Galton, F.: Problem 4001. Educational Times April 1 (17) (1873)
15. Galton, F., Watson, H.W.: On the probability of the extinction of the families. J. Royal Antropol. Soc., London 4, 138–144 (1874)
16. Garetto, M., Giaccone, P., Leonardi, E.: On the effectiveness of the 2-hop routing strategy in mobile ad hoc networks. In: Proceedings of ICC 2007 (2007)
17. Grishechkin, S.A.: On a relation between processor sharing queues and Crump-Mode-Jagers branching processes. Advances in Applied Probability 24, 653–698 (1992)
18. Helgason, O., Karlsson, G.: On the effect of cooperation in wireless content distribution. In: Proceedings of WONS 2008 (2008)
19. Heyde, C.C., Senneta, E.: Studies in the History of Probability and Statistics. XXXI. The simple branching process, a turning point test and a fundamental inequality: A historical note on I. J. Bienaymé, Biometrika 59(3), 680–683 (1972)
20. Key, E.: Limiting distributions and regeneration times for multitype branching processes with immigration in a random environment. Annals of Probability 15, 344–353 (1987)
21. Lambert, A.: The genealogy of continuous-state branching processes with immigration. Journal of Probability Theory and Related Fields 122(1), 42–70 (2002)

22. Le Gall, J.F.: Random trees and spatial branching processes. Maphysto Lecture Notes Series (Univ of Aarhus), vol. 9 (2000)
23. McNamara, J., Houston, A., Collins, E.: Optimality models in behavioral biology. SIAM Review 43(3), 413–466 (2001)
24. Núñez Queija, R.: Processor-Sharing Models for Integrated-Services Networks. PhD thesis, Eindhoven University of Technology (2000)
25. Resing, J.: Polling systems and multi-type branching processes. Queueing Systems 13, 409–426 (1993)
26. Roitershtein, A.: A note on multitype branching processes with immigration in a random environment. Annals of Probability 35(4), 1573–1592 (2007)
27. Watson, H.W.: Solution to problem 4001. Educational Times August 1, 115–116 (1873)

Routing in Quasi-deterministic Intermittently Connected Networks

Paolo Giaccone[1], David Hay[1], Giovanni Neglia[2], and Leonardo Rocha[2]

[1] Dipartimento di Elettronica, Politecnico di Torino, Italy
name.surname@polito.it
[2] Maestro project-team, INRIA Sophia-Antipolis Méditerranée, France
name.surname@sophia.inria.fr

Keywords: Intermittently Connected Networks, Delay Tolerant Networks, Routing.

Extended Abstract

Some of the recent applications using wireless communications (wildlife monitoring, inter-vehicles communication, battlefield communication,...) are characterized by challenging network scenarios. Most of the time there is not a complete path from a source to a destination (because the network is sparse), or such a path is highly unstable and may change or break while being discovered (because of nodes mobility and time-variations of the wireless channel). Networks under these conditions are usually referred to as Intermittently Connected Networks (ICNs) or Delay Tolerant Networks (DTNs). In such scenarios information delivery is then based on the store-carry-forward paradigm: a mobile node first stores the routing message from the source, carries it from a physical location to another and then forwards it to an intermediate node or to the destination. Typical examples of ICNs are those where nodes are intrinsically mobile (independently from data transfer purpose): vehicular networks [2] (in which data is carried over cars and buses), "pocket area networks" [2] (in which data is carried by people carrying small devices like PDAs), mixed ground/satellite networks and networks of sensors attached to animals [9]. Also some scenarios in which some nodes are mobile and some nodes are fixed (e.g, mobile devices with fixed gateways) present the same challenges.

The wide range of applications, promising performance results and concise modeling have led to an extensive research on ICNs during the last few years (e.g., [5,6,7,3]). At the core of this research line are *routing and scheduling algorithms*: at any given time, each node should find when and where to forward the data stored in its buffer so that it reaches the destination in a timely manner. Moreover routing for ICNs is not only limited to *forwarding schemes*, where a single copy of each packet is present in the network [1], but it also include *replication schemes*, which send many copies of the same data packet across the network. A prime example of replication schemes are *epidemic* routing

E. Altman et al. (Eds.): Bionetics 2009, LNICST 39, pp. 126–129, 2010.

algorithms (a.k.a *flooding algorithms*) in which each node sends each packet to *all* its neighbors. Replication improves performance in terms of delivery probability and delivery delay when contacts cannot be predicted or when transmissions are unreliable, but at the same time it implies higher costs in terms of required bandwidth, transmission energy and buffer requirements (see [10]).

Most of the research on routing in ICNs has focused on two extreme cases: 1) when contacts among nodes are deterministic and known in advance (e.g. in the case of space communications among satellites, probes and earth or space stations [8]) or 2) when they cannot be predicted (e.g. for human and animal mobility [2,9]) and are supposed to obey to some generic random mobility model, like random way-point, random direction or brownian models. Many interesting scenarios do not fall in any of these two cases: even complex mobility patterns often exhibit some form of periodicity or in other cases the underlying node mobility is known in advance, but it can be modified by random effects. A clear example is that of a vehicular network carrying data over public transportation (e.g., buses): the predictions of the contact times are derived from the schedule and routes of the buses; on the other hand, delays in bus operations clearly change the contact times or even prevent contacts to occur, implying that the predictions are not necessarily accurate.

Our preliminary investigation suggests that there is currently no framework to study comprehensively all the range of possible scenarios between deterministic contacts and unpredictable random contacts. For this reason, we have decided to investigate a specific class of networks characterized by *small deviations* from the deterministic contact model. We refer to such networks as "quasi-deterministic" ICNs.

Our current research consider bus networks as a case study for the general problem of routing in quasi-deterministic networks. Besides being an interesting application scenario itself, this specific network scenario will allow us to understand the key issues our models and our algorithms need to consider.

We envision that the infrastructure of bus-enabled data network is formed by (some) buses and bus stations equipped with wireless devices, e.g. based on WiFi technologies like in Dieselnet [2]. When two of them come within transmission range of each other, they can transfer data. Some access points at bus stops can also be connected to the Internet. Passengers on a bus (/waiting at a stop) can use this infrastructure through their mobile devices, associating to the bus (/stop) access point.

Here we focus on the simple case when we want to transfer some data from a bus or a bus stop to a remote bus or a remote stop. We envision two possible applications even for this simple unidirectional scheme. First, the data could be some non-time-critical information collected by sensors on the bus/stop for operation/management purpose that needs to be transfered to the bus system central operation point via a bus stop connected to the Internet. Second, the data could be destined to a passenger. We can think about possible hybrid systems, where, for example, the user requests its emails or a file through the standard cellular data connection and then get the reply through the DTN, that could offer a cheap data transfer service.

There are different options in designing such a system. First, the system could rely only on forwarding -i.e. a single copy of the data is propagated along a path- or could take advantage of multiple copies spread in the network to increase delivery probability and reduce delivery time. A second choice is between exploiting only transmission opportunities between buses and bus stops or exploiting also direct transmission opportunities between different buses. In the latter case we can expect the system capacity to increase, but at the same time meetings between buses are more unreliable, not only in terms of the time they are going to occur, but also in terms of their existence itself, being that delays can prevent buses to miss a meeting opportunity.

In order to gain a better feeling of reasonable modeling assumptions and of possible design choices we have started considering the actual public bus network in Turin, Italy, that has about 50 frequency based bus lines (up to 12 buses per hour) and 3000 bus stops. Our current contributions follow.

1. Analyzing real bus traces, we have characterized the statistical properties of bus delays at stops.
2. In [4], some of us have determined optimal routing schemes under deterministic contacts. We have then evaluated to which extent these schemes are robust to noise in the meeting process, i.e. how performance decrease when routing is based on predicted contact times ignoring the presence of noise.
3. Given the contact predictions and *a priori* statistical information on the noise process, we have developed a *multi-hop routing and scheduling algorithm* and evaluated its performance as a matter of throughput, delay and delivery probability.

References

1. Balasubramanian, A., Levine, B., Venkataramani, A.: DTN routing as a resource allocation problem. SIGCOMM Comput. Commun. Rev. 37(4), 373–384 (2007)
2. Burgess, J., Gallagher, B., Jensen, D., Levine, B.N.: MaxProp: Routing for vehicle-based disruption-tolerant networks. In: IEEE INFOCOM (2006)
3. Delay tolerant networking research group
4. Hay, D., Giaccone, P.: Optimal routing and scheduling for deterministic delay tolerant networks. In: Sixth International Conference on Wireless On-Demand Network Systems and Services, WONS 2009, February 2009, pp. 27–34 (2009)
5. Jain, S., Fall, K., Patra, R.: Routing in a delay tolerant network. In: ACM SIGCOMM, pp. 145–158 (2004)
6. Liu, C., Wu, J.: Scalable routing in delay tolerant networks. In: ACM MobiHoc, pp. 51–60 (2007)
7. Liu, C., Wu, J.: Routing in a cyclic mobispace. In: ACM MobiHoc, pp. 351–360 (2008)

8. Wood, L., Ivancic, W., Eddy, W., Stewart, D., Northam, J., Jackson, C., da Silva Curiel, A.: Use of the delay-tolerant networking bundle protocol from space. In: Proc. of the 59th International Astronautical Congress (September 2008)
9. Zhang, P., Sadler, C.M., Lyon, S.A., Martonosi, M.: Hardware design experiences in zebranet. In: SenSys 2004: Proceedings of the 2nd international conference on Embedded networked sensor systems, pp. 227–238. ACM, New York (2004)
10. Zhang, X., Neglia, G., Kurose, J., Towsley, D.: Performance modeling of epidemic routing. Comput. Netw. 51(10), 2867–2891 (2007)

Characteristics of the Dynamic of Mobile Networks[*]

Pierre Borgnat[1], Éric Fleury[2], Jean-Loup Guillaume[3], and Céline Robardet[4]

[1] Université de Lyon, ENS Lyon,
Laboratoire de Physique, UMR 5672 CNRS
69364 Lyon cedex, France
[2] Université de Lyon, ENS Lyon
INRIA/D-NET, LIP, UMR 5668 CNRS
69364 Lyon cedex, France
[3] Université Pierre & Marie Curie
LIP6 UMR 7606 CNRS
75252, Paris Cedex 05, France
[4] Université de Lyon, INSA-Lyon
LIRIS UMR 5205 CNRS,
69622 Villeurbanne cedex, France

Abstract. We propose in this paper a novel framework for the study of dynamic mobility networks. We address the characterization of dynamics by proposing an in-depth description and analysis of two real-world data sets. We show in particular that links creation and deletion processes are independent of other graph properties and that such networks exhibit a large number of possible configurations, from sparse to dense. From those observations, we propose simple yet very accurate models that allow to generate random mobility graphs with similar temporal behavior as the one observed in experimental data.

Keywords: Dynamic Networks, Network Models, Complex Systems, Random Graphs, Statistical Analysis, Stochastic Process, Data Mining.

1 Introduction

During the last decade, the study of large scale complex networks has attracted a substantial amount of attention and works from several domains: sociology [15], biology [7], computer science [1], epidemiology [12]. This emerging domain has proposed a large set of tools that can be used on any complex network in order to get a deep insight on its properties and to compare it to other networks. Such *fundamental* properties [1,10,11] are used as characterization parameters in the

---

[*] This work is partially financed by the European Commission under the Framework 6 HealthCare Project LSH PL037941 *"Mastering hOSpital Antimicrobial Resistance and its spread into the community"* (MOSAR) and AEOLUS project IST IP-FP6-015964. The views given herein represent those of the authors and may not necessarily be representative of the views of the project consortium as a whole.

E. Altman et al. (Eds.): Bionetics 2009, LNICST 39, pp. 130–139, 2010.

study of various problems such as virus spreading [5,9,12] in the epidemiology context, or information / innovation diffusion [2,6] for instance. Whereas most of such complex networks are inherently dynamic, this aspect has less been studied. Most approaches consider *growing models*, such as the preferential attachment model [2,8] or analyze the aggregation of all interactions. Both approaches may miss the real dynamic behavior while there is a strong need for dynamic network models in order to sustain protocol performance evaluations and fundamental analyzes.

In this paper, we address the description and the simulation of sensor mobility networks. The proposed methods come from various research domains (signal processing, graph theory and data mining). This emphasizes the necessity of interdisciplinary research since dynamic networks are becoming a central point of interest, not only for engineers and computer scientists but also for people in many other fields.

We apply those methods on mobility networks. Mobile devices with wireless capabilities are a typical example of evolving networks where users are spread in the environment and communications can only take place if they are near each others. We study an empirical mobility network, called IMOTE [3], based on 41 Bluetooth sensors whose interactions have been recorded during 3 days. This who-is-near-whom network evolves every time users move.

We introduce some simple methods to describe the network dynamics and propose models of dynamic networks. The complete methodology of analysis was reported in a full version of this communication [14].

1. We study graph properties as function of time to provide an empirical statistical characterization of the dynamics.
2. We also compute global indicators from the dynamics of the network (connected components, triangles, and communities).
3. We propose models to perform random dynamic networks simulations.

The descriptive analysis show that link (or edge) creation and deletion processes is mostly independent of other graph properties and that such networks exhibit a large number of possible configurations, from sparse to dense. From those observations, we propose simple yet accurate models that allow to generate random mobility graphs with similar temporal behavior as the one observed in experimental data.

Even though such networks have obvious specificities, the in-depth study of their dynamic is an original work, and can have a broader impact on the complex system community. It is noteworthy that our approach does not make any assumption on the specificities unlike agent-based models or geographical approaches.

2   Statistical Analysis of Snapshots of Graphs

We first propose and study a set properties usable as a practical basis for the analysis of dynamic mobility networks that can be easily extended to large complex networks. The studied graph properties are the distributions of contact and inter-contact durations, the correlation between various graph properties as function of time and the links correlation, so as to give an empirical statistical characterization of the dynamics.

2.1   Contact and Inter-contact Durations

The contact and inter-contact duration distributions are dynamic characteristics that are interesting for mobility networks. The contact duration is the time during which two vertices remain directly and continuously adjacent. The inter-contact duration is the duration between two periods of contact for two vertices.

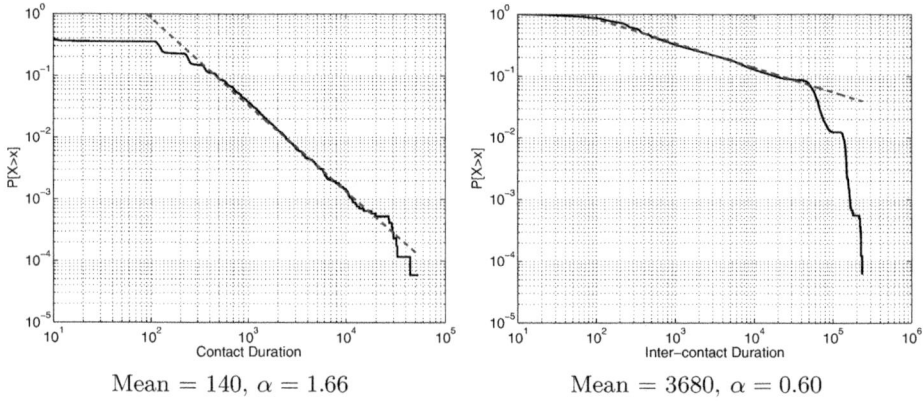

Mean = 140, $\alpha = 1.66$ Mean = 3680, $\alpha = 0.60$

Fig. 1. Contact (left) and Inter-contact (right) duration distributions (CCDF)

Fig. 1 shows that the contact and inter-contact durations have a power-law behavior [4] (mean and estimated α are reported in captions). Inter-contact duration distribution has a very strong variability due to long periods of lack of contact for some nodes, whereas the distribution of contact durations is less heavy-tailed. The heavy-tailed nature of these distributions seems to be an ubiquitous property of dynamic mobility networks.

2.2   Correlation of Graph Properties

We compute in Table 1 the correlation coefficient between several graph properties seen as functions time t: $E(t)$ is the number of active links, $V(t)$ is the number of connected vertices, $N_c(t)$ is the number of connected components, $D(t)$ is the average degree, $T(t)$ is the number of triangles, $E_\oplus(t)$ is the number of links added at time t and $E_\ominus(t)$ is the number of links removed at that time.

Most of the correlation coefficients are rather high. This is mostly due to the fact that there are constraints on the properties of graphs. For instance

Table 1. Correlation coefficients between various graph properties studied as functions of time

	$E(t)$	$V(t)$	$N_c(t)$	$D(t)$	$T(t)$	$E_\oplus(t)$	$E_\ominus(t)$
$E(t)$	1	0.85	-0.56	0.95	0.90	**0.19**	**0.15**
$V(t)$	0.85	1	-0.20	0.70	0.66	**0.15**	**0.11**
$N_c(t)$	-0.56	-0.20	1	-0.70	-0.41	**-0.16**	**-0.15**
$D(t)$	0.95	0.69	-0.69	1	0.86	**0.19**	**0.15**
$T(t)$	0.90	0.66	-0.41	0.86	1	**0.15**	**0.11**
$E_\oplus(t)$	**0.19**	**0.15**	**-0.16**	**0.20**	**0.15**	1	0.03
$E_\ominus(t)$	**0.15**	**0.11**	**-0.15**	**0.16**	**0.10**	0.03	1

the number of links $E(t)$ has a strong influence on the number of connected vertices $V(t)$. Furthermore, the time series are not stationary and there are clear periods of one day and variations between days and nights. Only link creation and deletion processes ($E_\oplus(t)$ and $E_\ominus(t)$) remain mostly uncorrelated with all other properties. Their evolution can be considered mostly independent from the one of other graph properties.

2.3   Links Correlations

Let us now turn to individual links. The correlation coefficient of the state evolution of links characterizes the dependency between links. The state evolution $S_e(t)$ of each link e is equal to 1 if link e is in the mobility graph at time t and 0 otherwise. The correlation matrix $Co(e, e')$ for links is computed as:

$$Co(e, e') = CORR(S_e, S_{e'})$$
$$= < S_e(t)S_{e'}(t) >_t - < S_e(t) >_t < S_{e'}(t) >_t \ .$$

For each link, we also compute its average correlation coefficient with respect to the other links as the average of absolute values. This helps to keep track of the strength of the correlation rather than its direction.

Fig. 2 shows the histogram of the values. Most pairs of links have a very low correlation coefficient. Rare couples of links exhibit a strong correlation.

2.4   Joint Distribution

The empirical joint distribution of the number of connected nodes and the number of links gives a finer description of the dependencies between those two properties.

As expected, the correlation between vertices and links shown on Fig. 3 is positive: the more vertices are connected, the more links are present. However it is worth noting that the variation of the number of links is not constant over the number of vertices. For a given number of vertices, the network can have a large number of possible configurations, some of which are very sparse and others more dense, as shown by the gray scale in the plots.

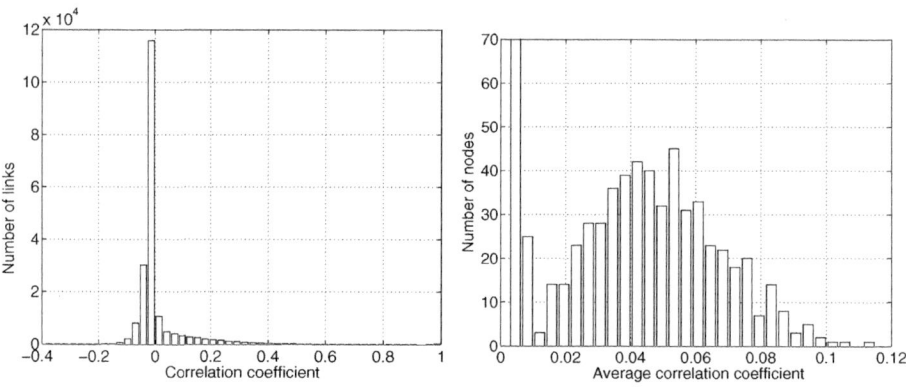

Fig. 2. Correlation between links (left) and average correlation for each node (right)

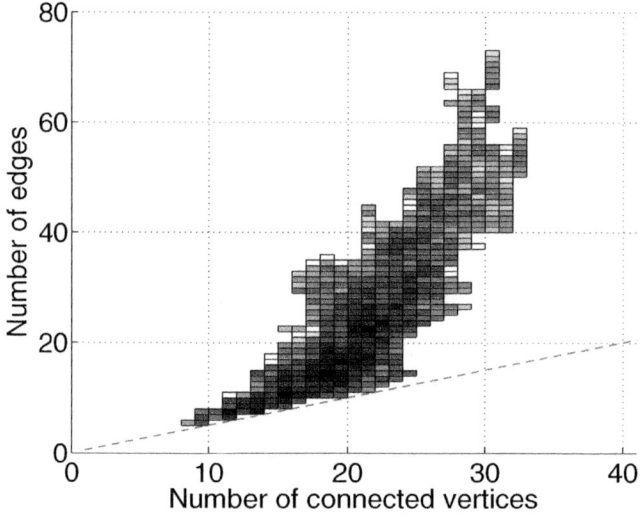

Fig. 3. Joint distribution of the number of connected nodes and the number of links. The gray scale is proportional to the logarithm of the probability (darker means higher probability, and white no occurrence of this event in the data).

3   Global View of the Dynamics

We also propose to study global indicators from the dynamics of the network (stability of connected components, triangles creations, existence of communities).

3.1   Triangles

The existence and persistence of connected components is generally associated with a rather large number of triangles in the graph. Therefore, an important characteristic of the dynamic is the evolution of the number of triangles in time. To evaluate the proportion of links that create triangles when they appear, we compute the number of link creations that leads to an increase of the number of triangles in the graph or that does not change it.

Table 2. Proportion $P_{+/tri+}$ (resp. $P_{+/tri=}$)) of links creations that add new triangles (resp. not), and the average proportion $f_{+/tri+}(resp.f_{+/tri=})$) of inactive links that, if created, would add a triangle, (resp. not).

	$P_{+/tri+}$	$P_{+/tri=}$	$f_{+/tri+}$	$f_{+/tri=}$
IMOTE	44 %	56 %	6 %	94 %
RANDOM	10 %	90 %	5 %	95 %

These proportions are given in Table 2. Around 40% of links creations increase the number of triangles in the graph whereas this proportion equals 10% in a random (Erdös-Rényi) graph with the same numbers of vertices and links. The proportion of inactive links that would create a triangle is very low for IMOTE data set and the simple random graph. This emphasizes the fact that this is not because more links can create triangles that the proportion P+/tri+ is higher in experimental data: it is on the contrary an intrinsic property of the dynamics. As the proportions of links that could create a triangle are similar in both graphs $(f_{+/tri+})$, this phenomenon is characteristic of real graphs: links creations tend to create triangles in fairly large proportion.

3.2   Dynamic Communities

To describe the graph structure evolution, we isolate "communities", which are commonly considered as large groups of individuals who interact intensively with each other over a long period of time. A community can be seen as a dense connected sub-graph that appears in a large number of time steps (not necessarily consecutive).

We compute the set of connected sub-graphs having more than σ links and that are included in at least τ graphs:

$\mathcal{C} = \{S = (V, E), |\{t \mid S \subseteq G_t\}| \geq \tau$ and $|E| \geq \sigma$ and S is connected$\}$. Then, the denser sub-graphs are selected using a density threshold that selects the most important and established ones. Finally, the trajectories of individuals among social groups are inferred: an arc (u,v) represents individuals moving at least once in the data from group u to group v. Fig. 4 shows the identified communities and their dynamic. For example, individual 8 initially belongs to group 13, he/she further moves into group 6, and finally enters group 7.

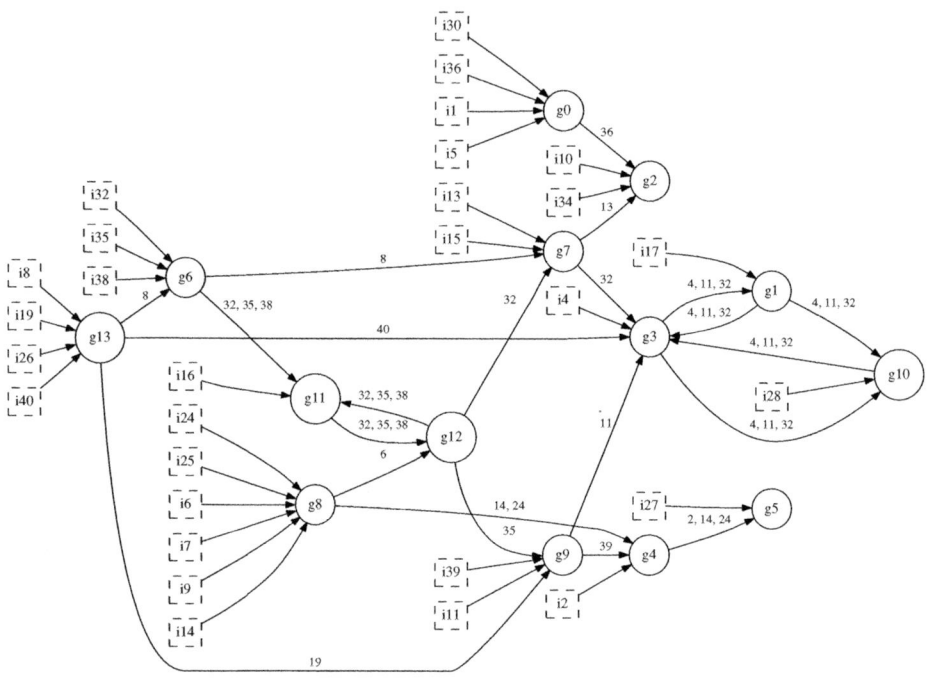

Fig. 4. Individual trajectories in groups ordered by time. i_x (boxes) are individuals while g_x (circles) denotes social groups.

4   Modeling of the Dynamics

From the previous set of analysis, we propose generic random dynamic models that allows to generate random dynamic graphs which have a behavior similar to the one observed in experimental data set. Their design is justified by the previous observations. First, as the contact and inter-contact are power-law distributed (as seen in Section 2.1), those non-trivial empirical distributions should be taken into account when constructing a model of the data. Second, computation of empirical times of correlations show that link creation/removal process is less correlated in time that other graphs properties seen in Sect. 2.2. It is also reasonable to consider that links evolutions are uncorrelated in time. These characteristics justify the use of a simple Markovian (memory-less) link creation/removal process. Finally, we observed that for a given number of nodes, the network can have a large number of possible configurations, some of which are very sparse and others more dense. Thus our model should be able to reproduce this property.

The simulation is based on a transition model with Markovian property. For each time step and for each link independently, each link changes its state (active or inactive) using a transition probability depending on the time since the

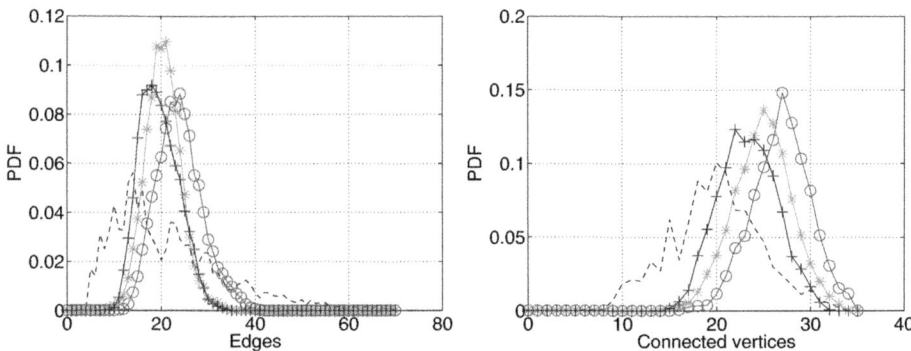

Fig. 5. Probability distribution function (PDF) for original data and the classical models. On the left we plot the PDF for the number of connected vertices and on the rigth we plot the number of edges. The plots are for original data (−) and for several models: imposing the sole contact and inter-contact duration distribution (-o- on the plot), or adding the statistics of N_C (-*-) or V (-+-).

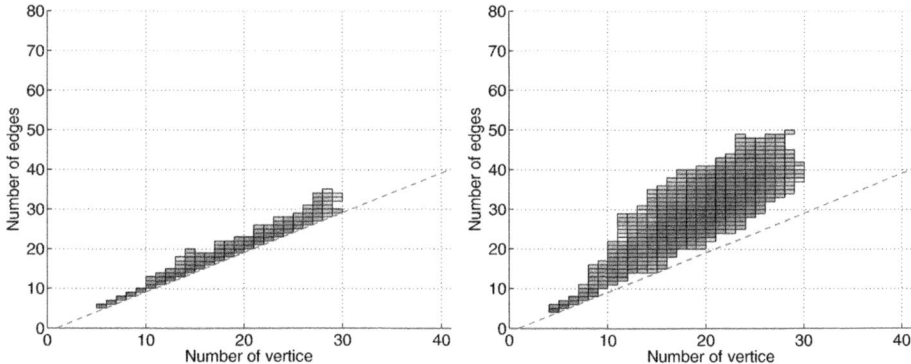

Fig. 6. Probability distribution function (PDF) for original data and the classical models. On the left and right we plot joint distribution of the number of connected vertices and links in connected components: a model imposing the contact and inter-contact duration distribution and the statistics of N_C (middle plot) fails at reproducing the correct behaviour; a model respecting also the statistics of the triangle dynamics (right plot) reproduces much more the empirical behaviour of the data.

link is in its current state. In addition, the probability of transition is weighted by a probability of acceptance of the new state depending of the experimental distribution for a property of interest such as $E(t)$, $V(t)$, $N_C(t)$ and $D(t)$. This is implemented by Rejection Sampling [13], based on a Metropolis-Hasting algorithm and take into account the target distribution of the graph property. To take into account in the simulations that the average proportion of link creations that yield triangles is larger than for random graphs, a weight is applied on the transition probability to reproduce the correct dynamical transition process concerning triangles.

In order to justify that the use of properties (like $E(t)$, $V(t)$, $N_C(t)$) within the model in addition to the sole contact and inter contact duration distribution we compare such distribution with the original Data set. A first remark is that the sole contact and inter-contact duration distributions dramatically fail to reproduce the properties. More precisely, the number of connected vertices is strongly over-estimated, the number of connected components is under-estimated, and so is the number of triangles. The non-stationarity in the IMOTE data introduces a much higher variance, yet it does not explain all the differences.

Fig. 5 illustrates the way the models work. It is based on the distribution of $E(t)$ and $N_C(t)$ and produces qualitatively similar characteristics than the real data set. The probability distribution functions of $N_C(t)$ (on the rightmost figure) for the original data (black), and the different models proposed are shown. We can observe that by imposing the distribution of the number of connected vertices improves the accuracy of the simulations.

However, when analyzing more refined characteristics such as the joint distribution of number of links and vertices in connected components, the original data set (shown on Fig 3) and the simulations of the models (shown on Fig. 6) are much different even when we impose number of connected vertices distribution. The connected components in models (sole inter contact and/or with the introduction of graph property PDF) are much less dense. We believe this is of major importance for communication protocol design and realistic models have to reproduce this property and this yield to the introduction of the number of triangle property (the results is depicted on the rightmost figure of Fig 6). This time, the density of connected components is comparable to the original data set. Note that introducing the dynamic of triangles also yield to the creation of "social" groups like the ones depicted in Fig. 4. When using the classical models, the graph is too sparse and the community phenomena is not reproduced in the model.

Our investigations have shown that the model, thanks to the introduction of dynamical characteristics such as the evolution of the number of triangles, manages to generate more realistic simulations. This opens the track to improved models that match the important characteristics of dynamics of mobility networks. This is illustrated by the two last figures that show the dynamic communities in the output of the simulation and the joint probabilities of the number of connected vertices and links in the graph.

This study opens the track to improved models that match the important characteristics of dynamics of mobility networks.

5   Conclusion

By introducing several models, we are able to highlight the diversity of properties that are needed to characterize such networks. Furthermore, our models provide insight into existing notions of dynamic networks and demonstrate that the structure and the dynamics are complex and are not a direct consequence of the contact and inter-contact durations. Proposing such models is crucial since it enables a validation of the ongoing research conducted in the various areas that deal with dynamic networks. It has also many applications in performance evaluation for instance.

References

1. Albert, R., Barabasi, A.-L.: Statistical mechanics of complex networks. Reviews of Modern Physics 74, 47 (2002)
2. Albert, R., Jeong, H., Barabasi, A.: The diameter of the World Wide Web. Nature 401, 130–131 (1999)
3. Chaintreau, A., Crowcroft, J., Diot, C., Gass, R., Hui, P., Scott, J.: Pocket switched networks and the consequences of human mobility in conference environments. In: WDTN, pp. 244–251 (2005)
4. Chaintreau, A., Crowcroft, J., Diot, C., Gass, R., Hui, P., Scott, J.: Impact of human mobility on the design of opportunistic forwarding algorithms. In: INFOCOM (2006)
5. Ganesh, A., Massoulie, L., Towsley, D.: The effect of network topology on the spread of epidemics. In: INFOCOM (2005)
6. Gruhl, D., Liben-Nowell, D., Guha, R.V., Tomkins, A.: Information diffusion through blogspace. SIGKDD Explorations 6(2), 43–52 (2004)
7. Kempe, D., Kleinberg, J., Tardos, E.: Maximizing the spread of influence through a social network. In: KDD, pp. 137–146. ACM Press, New York (2003)
8. Leskovec, J., Chakrabarti, D., Kleinberg, J., Faloutsos, C.: Realistic, mathematically tractable graph generation and evolution, using kronecker multiplication. In: Jorge, A.M., Torgo, L., Brazdil, P.B., Camacho, R., Gama, J. (eds.) PKDD 2005. LNCS (LNAI), vol. 3721, pp. 133–145. Springer, Heidelberg (2005)
9. Meyers, L., Pourbohloul, B., Newman, M., Skowronski, D., Brunham, R.: Network theory and sars: Predicting outbreak diversity. J. Theor. Biol. 232, 71–81 (2005)
10. Newman, M.: The structure and function of complex networks. SIAM Review, 167–256 (2003)
11. Newman, M.E., Park, J.: Why social networks are different from other types of networks. Phys. Rev. E Stat. Nonlin. Soft. Matter Phys. 68, 036122 (2003)
12. Pastor-Satorras, R., Vespignani, A.: Epidemic spreading in scale-free networks. Phys. Rev. Let. 86, 3200–3203 (2001)
13. Robert, C., Casella, G.: Monte Carlo Statistical Methods. Springer, Heidelberg (2004)
14. Scherrer, A., Borgnat, P., Fleury, E., Guillaume, J.-L., Robardet, C.: Description and simulation of dynamic mobility networks. Computer Networks 52(15), 2842–2858 (2008)
15. Watts, D., Strogatz, S.: Collective dynamics of small-world networks. Nature 293, 420–442 (1998)

ONTO-RUP: A RUP Based Approach for Developing Ontogenetic Software Systems

Farida Kherissi and Djamel Meslati

LRI Laboratory,
University Badji Mokhtar-Annaba
BP 12, 23000, Annaba, Algeria
Kherissi_farida@hotmail.com, Meslati_djamel@yahoo.com

Abstract. It is impossible to produce systems of any size which do not need to be changed. Once software is put into use, new requirements emerge and existing requirements change as the business running that software changes. Ontogenetic software systems have the ability to evolve dynamically in an autonomous way to meet the user needs and the anticipated and unanticipated changes of requirements. The evolution of these systems has the particularity to be a continuous process that shapes them from the beginning of their creation. This characteristic does not match the current development methods which consider the evolution a sporadic process. All current methods are still unsuitable for development of ontogenetic software systems. Indeed, they do not provide any tool or artifact to take into account the anticipated and unanticipated changes. In this article, we propose an extension of the Rational Unified Process that aims at providing a preliminary framework that allows developing ontogenetic systems.

Keywords: Use Cases, Change Cases, Ontogenetic Systems, Rational Unified Process.

1 Introduction

Many of IT experts talk about evolution modeling and/or developing flexible software [1, 6, 8, 11]. They consider flexibility a good feature and ease of software system modification one of the most important attributes, but they do not say how to achieve it. The knowledge of a system's requirements is necessary imperfect because a significant part of those requirements lies in the future and is unknowable at the time the system is designed. The software engineering community expects that most of the future innovations in information technologies will likely take place in the context of the development methods. These methods have to face the software evolution problems and complexity. Indeed, we are now approaching the limits of our capabilities to deal with complex software systems [17]. Inspired by nature and biology, researchers are now considering alternative solutions to software evolution problems such as ontogenetic software systems, where the main feature is to evolve dynamically and in an autonomous way. Like biological organisms ontogenetic systems

E. Altman et al. (Eds.): Bionetics 2009, LNICST 39, pp. 140–151, 2010.

evolve dynamically and the change mechanisms shaping them throughout their lives are considered as their ontogenetic imension (also called ontogenesis). Taking into account ontogenesis, when developing software systems, is a new challenge and a radical vision of the evolution that will have an influence on our perception of software systems as well as on our approaches of designing them. As an illustrative example, Mage is a bio-inspired approach of ontogenetic systems that is based on genetics [14, 15]. It provides concepts to model the ontogenesis of a software as an embedded genome, whose role consists of shaping continuously this system according to the anticipated and unanticipated changes that may occur. While anticipated changes are those identified before the release of the software system, unanticipated changes emerge when the system is used. According to the Mage approach, a system is composed of two parts a genome and a phenotype. The phenotype is the equivalent of the classical software code. The genome is a collection of genes which continuously shape the phenotype.

The Mage approach needs and advocates devising new suitable development methods in order to benefit from the ontogenetic systems to the full extent. In this context, extending the existing methods is a promising approach.

Although RUP is based on a set of engineering best practices, it is not adapted for ontogenetic software systems. Indeed, RUP deals with changes of requirements only before the software release. Anticipating future changes and building a system to change is not an option in RUP.

In this article, we propose ONTO-RUP as an extension that adapts RUP to the development of the ontogenetic systems and especially Mage-like systems which support a modeling of the changes in the form of a genome that shapes the system in an autonomous way. We are interested, here, in the key phase related to the definition and analysis of the requirements' evolution.

In the remainder of this article, we present briefly, in section 2, limitation of current development methodologies, then, we present the proposed approach ONTO-RUP. Section 5 shows how the extension works with an example. Section 6 compares this work with related ones. Finally, section 7 gives a conclusion and enumerates some research issues.

2   Limitation of Current Development Methodologies

2.1   Examples in Real World

Real world software systems need to evolve continually in order to cope with ever-changing user's requirements. For example, learning software systems have to evolve autonomously and dynamically to deal with evolutions affecting the software after its release, such as:

- Evolution of system's functions, considering anticipated as well as unanticipated ones

- Anticipated evolution consisting of interface evolutions corresponding to several factors (pedagogical, technological, etc.)

- Unanticipated changes for updating some components without stopping the software system.

Another example is the banking transactions management system where evolutions are frequent:

- Anticipated evolution to deal with business rules such as allowing cash withdrawal for some customers even if the account balance becomes negative.

- Unanticipated evolution consisting of new authentication procedures such as using biometric identification devices instead of a credit card.

Ontogenetic systems clearly require suitable development process that deal with evolution as a fundamental concept. A development process for ontogenetic software systems must be provided with mechanisms to:

- **Capture evolution:** A specific process of eliciting anticipated and unanticipated changes that software systems undergo.

- **Modeling evolution:** Describing, structuring, and documenting changes.

- **Deal with implementation:** Including artifacts for coding anticipated changes and providing tools for generating interfaces, assisting unanticipated changes integration within the software code. For example, in the Mage approach, running software consist of a phenotype and a genome. The phenotype corresponds to the traditional code and the genome is composed of elements which continuously and dynamically shape the phenotype. The internal evolutions must be coded in the phenotype using the conventional if-then-else statements which change the system behavior according to some conditions or events.

2.2   Extending the Suitable Method

Methods are organized ways to produce software. They include several steps to follow during the development process, specific representations (graphical or textual), rules governing the system description and design guidelines [18]. In the software engineering domain, there are a number of development methodologies that have been adopted and/or successfully adapted to meet specific business needs. These range from traditional waterfall development to more recent ones like the rational unified process (RUP), and many agile development methods. In our context, rather than proposing a new development process from scratch, we prefer extending existing ones to deal with ontogenetic software development. We choose the RUP process for many reasons:

- RUP is use-case driven which is suitable for us as we propose (later in this article) a modified form of use cases [3, 16] for modeling changes.

- RUP promote the participative development. This is a key feature to deal with cooperative understanding of requirement evolutions.

- RUP is founded by best practices.

2.3   Rational Unified Process

Rational Unified Process is an approach for developing software, which is iterative, architecture-centric, and use-case driven [10]. The RUP is a well-defined and well-structured software engineering process. It clearly defines project milestones, who is responsible for what, how things are done, and when they should be done. The RUP is structured on two axes or dimensions: The dynamic aspect (horizontal) expresses cycles, phases, iterations, and milestones; the static aspect (vertical) expresses activities, disciplines, artifacts, and roles (figure1.).

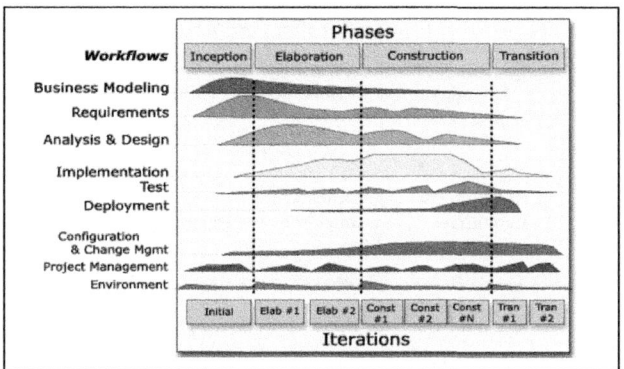

Fig. 1. Dimensions of RUP [10]

2.4   Insufficiencies of the RUP

What is significant in the RUP methodology compared to the others ones is the recognition of the primacy of changes even late in the development cycles compared to the traditional one with the aim of making software development more predictable and more efficient. The ontogenesis implementation requires tools and artifacts to design software system for changing after product release. Unfortunately, RUP and all current development methodologies fall short to deal with ontogenetic software systems.

3   Proposed Solution

A software development process describes *who* is doing *what, how*, and *when* [7]; presented in RUP terminology respectively by, workers, activities, artifacts, and workflows. Its adaptation for the ontogenetic systems is relatively a promising and effective approach. We propose an extension called ONTO-RUP which we describe in what follows.

3.1   A New Structure to the RUP

The ONTO-RUP includes, in addition to the usual phases of RUP (i.e. inception, elaboration, construction, and transition), a new phase we have called "Evolution" that takes place between inception and elaboration phases (Figure 2). The position of the evolution phase in the life cycle corresponds to the level of the progression of the discipline Requirements (Figure1.), we can see that a large portion of Requirements takes place in Inception, although it does continue through to early Transition. In Figure 1, we can observe that at this level, the functional requirements of the system are not all specified (about 20% of the use case model). But, this portion of use case models describes the overall system's behavior which are critical since they have the most important influence on the system's architecture and design.

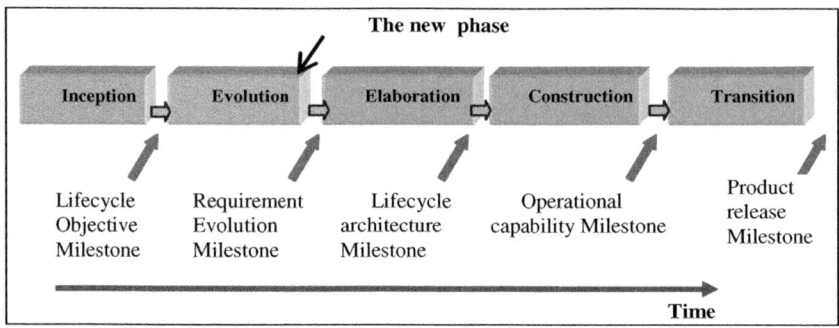

Fig. 2. Phases of the ONTO-RUP

Each phase is concluded with a well-defined milestone; a point in time at which certain critical decisions must be made and therefore key goals must have been achieved. The "Requirement Evolution" milestone supposes that anticipated change cases model is specified in an overall way. During **Evolution phase** a large portion of proposed activities in this paper will occur.

3.2 Disciplines for Modeling Ontogenesis

RUP is organized around nine disciplines (Figure 1, vertical axis). A discipline is a collection of activities that are related to a major "area of concern" within the overall project. Disciplines group activities logically. ONTO-RUP proposes a set of disciplines:

- Anticipated evolutions requirement Discipline.

- Evolution Decomposition Discipline.

- Unanticipated evolution Discipline

Our focus, in this paper, is on the "Anticipated evolutions discipline". The purpose of this discipline is to:

- Establish and maintain agreement with the customers and other stakeholders on what the future system should do
- Provide system developers with a better understanding of the system requirements evolution

- Define the boundaries of (delimit) the future system
- Provide a basis for planning the technical contents of iterations.

3.3 Decomposition of Changes

Decomposition of changes deals with anticipated and unanticipated evolutions, and takes into account, for anticipated evolutions, internal ones (built-in the code of the software) and external ones (applied one the software during its execution when some conditions are met). Since anticipated evolutions may be internal or external and this may have an influence on the system performances, it is important to decide which evolution is internal and which one is external. To help developers in achieving this decision, we propose a discipline:"Decomposition Discipline".

4 The "Evolution Phase"

4.1 Objectives

The Evolution phase is the second phase after the inception phase in the life cycle of Onto-RUP. This phase is the backbone of the life cycle, because it deals with the ontogenetic dimension of the system. According to RUP, the inception phase is about understanding the project scope and objectives and getting enough information to confirm that we should/shouldn't proceed with the project. The Evolution phase will take place as the second phase during which there will be a study of the needs for evolution to give the comprehensive view of all aspects related to the evolution of the future system. The Evolution phase milestone presents a well-defined set of objectives. We enumerate the objectives of this phase:

- Understand the overall system evolution
- Eliciting evolution of requirements and elaborating the change case model (a modified form or use case that describes a change)
- Deciding if changes will be internal or external evolutions.

4.2 Inputs/Outputs of Evolution Phase

Figure 3 shows the input and the output of the evolution phase. In this phase change case model is globally elaborated and decisions are made about the nature of the major requirements evolution.

Among the objectives of the RUP inception phase we find the comprehension of the system to be built and the identification of the system fundamental functionalities [9, 10].

The use case model adopted by RUP presents the formal specification of requirements. This model separates the system in actors and use cases.

The Requirement discipline provides the system developers with a better understanding of the system requirements, delimits the boundaries of the system, provides a basis for planning the technical contents of iterations, provides a basis for estimating cost and time to develop the system, etc.

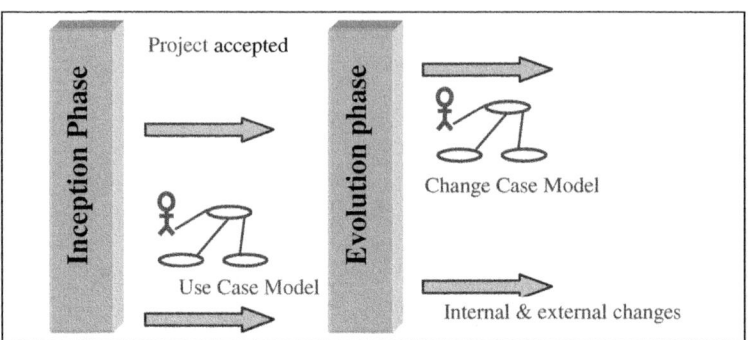

Fig. 3. Inputs/Outputs of the Evolution Phase

Use cases. They are specification of a sequence of actions and variants that a system (or other entity) can perform when interacting with external actors. Each use case describes the behavior of the system under various conditions and shows how it responds to a particular request from an external actor (called the *primary actor*). The primary actor initiates an interaction with the system to accomplish some goal, and the use case is a black box which describes the reaction of the overall system, without indicating how the system achieves the goal.

A complete set of use cases specifies all the possible use of a system and, consequently, describe all the necessary behavior of the system. In our approach, we use the use case formalism given by Cockburn in [2].

4.3   Activities of Anticipated Evolutions Requirement Discipline

Starting from use cases model elaborated in the inception phase, the purpose of the "Anticipated evolutions requirement Discipline" is to built the change cases model starting by use case models. These activities require an iterative process, where each performed iteration represents a unit of change which consists of one or more change cases. A unit of change will be an input for the sub-process called *changes study* using techniques for eliciting evolution/change requirements. We find in [13], a set of techniques for eliciting requirements. There are many other techniques and their combination seems to provide richer and more detailed requirements for evolution of requirements. We propose the following combination: begin with *questionnaires*, followed by *structured interview* to gain deeper insight into the possible evolution, then the *brainstorming* technique is used, and at the end the *group work* technique. his ast technique is appropriate in our context, because it suggest collaborative meeting that involve and commit the stakeholders directly and promote cooperation.

Figure 4 shows the steps composing the evolution phase. In the change study step we need to decide if the unit of change is further considered or ignored. In the next step, we extract the change cases model using the change case artifact we have proposed (described next). This formalism is based on the use case one but contains specific parts to deal with evolution.

Change cases. The change case is a powerful formalism to capture potential changes that has been first proposed by Ecklund in [3]. In our work we have introduced specific parts to deal with new requirements but also to specify modification to the already implemented ones. Change cases are characterized by both their simplicity and expressive power. Like the use cases, they concern the system in the whole.

4.4 Planification of the Evolution Phase

\In the evolution phase the first iteration aims at specifying the change cases in a overall way.

Those which follow will accentuate in studying the impact of the specified changes. The Requirements and Ontogenesis disciplines are related to other process disciplines: Analyze and Design, Test, Configuration and Change Management, and Project Management. The use case model, change case model, and Requirements Management Plan are important inputs to the iteration planning activities. Since a use/change case describes how a user will interact with the system, the use of UML notation [16] such as sequence diagrams or collaboration diagrams to show how this interaction will be implemented by the design elements. We can also identify test cases from a use/change case. This ensures that the services the users expect from the system are really provided.

5   Illustrative Example

One of the possible ways to evaluate software development process models or methodologies is to choose some exemplar systems as case studies and employ the process model or methodology in developing case study systems. Let us consider the banking transactions system previously introduced. In the following we describe the two phases of ONTO-RUP.

Inception Phase. In ONTO-RUP the Inception phase is typically equivalent to RUP approach. It provides rich set of activities and guidelines.

At the end of the inception phase, if there is an agreement about the project (does or not pass this milestone "Lifecycle Objective"), it can either be cancelled or it can repeat this phase after being redesigned to better meet the criteria. The resulting artifact is a model of use cases which presents an overall schema of current requirements (Figure 5).

Evolution Phase. The analysis of the change of requirements is held during the evolution phase using the change process. After the study of the possible anticipated changes using the techniques proposed in the previous section; decisions are taken concerning the anticipated changes that may affect the software system. Figure 7 summarizes all the use cases composing the global view of the future system. We use

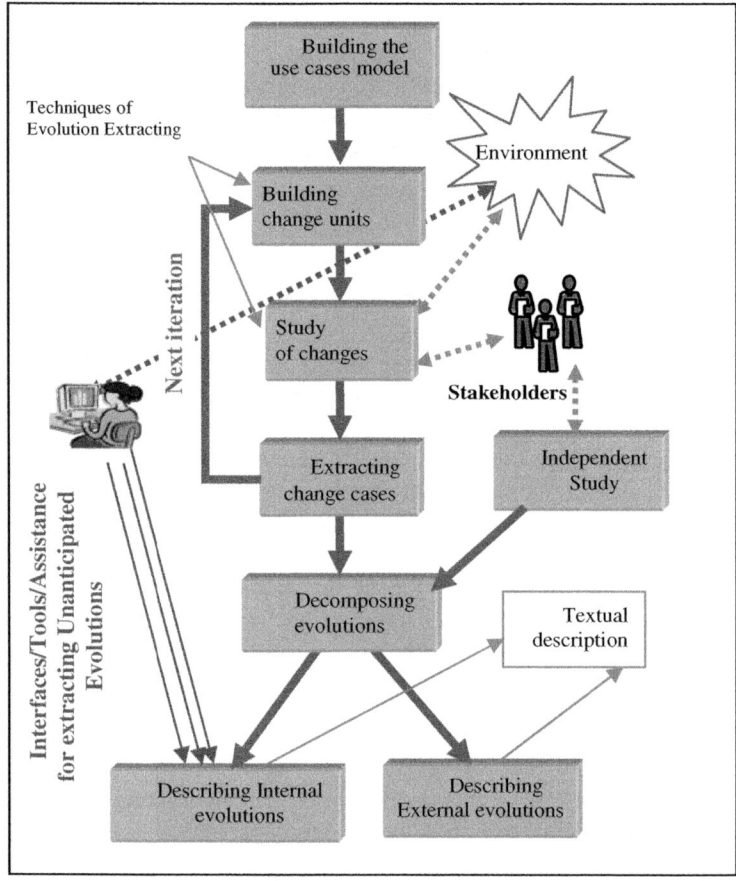

Fig. 4. Requirements evolution Process

the change case formalism suggested in the previous section for describing the anticipated changes. We give two change cases modeled for this system:

- Change case for adding an authentication procedure.

Change case for allowing some customers to make cash withdrawal even with a negative account balance.

- The second change case is an unanticipated change, that's why it doesn't appear in Figure 6. Notice that the change process remains the same for unanticipated changes.

Only few works deal with evolution as an important challenge. Mage considers the
evolution as a fundamental process and provides several concepts to capture and
model changes as a dynamic and autonomous process [14, 15]. The autonomic computing approach seeks to provide systems with aptitudes to be self-managed and self-
repaired [17]. To our best knowledge there is no complete development approaches
dedicated to autonomic computing. The same is true for

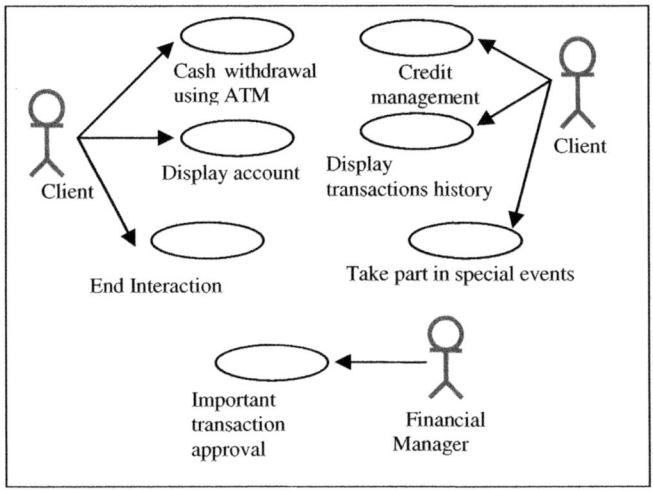

Fig. 5. Overall model of banking transaction system

Use case	Type of change	Description of Change
ATM cash withdrawal	Change	New authentication procedure
Display account	Maintain	No need of change for this use case
Request Credit	Maintain	No need of change for this use case
Credit management	Change	A potential client will benefit of an important credit
Important transaction approval	Maintain	No need of change for this use case
ATM cash withdrawal	Change	Changing the maximal of amount withdrawal of X every 3 years
Fidélisation Order 1	Change	Adding unit of fidelisation If date-date create=1
Fedilisation Order 2	Change	To encourage using ATM benefit> Y
ATM cash withdrawal	Change	Permit supply credit in 15/12 until 31/12 (End year)

Fig. 6. Studied use cases

6 Related Work

Only few works deal with evolution as an important challenge. Mage considers the evolution as a fundamental process and provides several concepts to capture and model changes as a dynamic and autonomous process [14, 15]. The autonomic computing approach seeks to provide systems with aptitudes to be self-managed and self-repaired [17]. To our best knowledge there is no complete development approaches dedicated to autonomic computing. The same is true for biomorphic systems described in [12]. The work in [4], [5] proposes an approach with an evolution process at the requirement level that uses the concept of gap. This approach is based on a meta-model and a generic typology of operators to express different kinds of evolution.

7 Conclusion

The ontogenetic software systems are particular systems which allow dynamic evolution of requirements. Motivations for research in this context are important and investigations are widely justified from the industrial and economic point of view since at least 50% of software costs are relative to their evolution. However, today, there are no development methods to support such systems.

We presented in this article an attempt in this direction, it proposes an extension of RUP which preserves its features while providing artifacts and methods to support the development of the ontogenetic systems.

As a perspective to this work, we need first to describe other disciplines related to *Modeling Ontogenesis Disciplines*. Another important perspective is to consider the evolutions in subsequent phases of the development process.

References

1. Chapin, N., Hale, J.E., Khan, K.M., Ramil, J.F., Tan, W.G.: Types of software evolution and software maintenance. J. Software Maintenance and Evolution: Research and Practice 13, 3–30 (2001)
2. Cockburn, A.: Writing Effective Use Cases. Addison Wesley, New York (2001)
3. Ecklund, E.F., et al.: Change Cases: Uses Cases that Identify Future Requirements. In: Proceeding OOPSLA 1996. ACM Press, New York (1996)
4. Etien, A., et al.: Overview of Gap-driven Evolution Process. In: AWRE 2004 9th Australian Workshop on Requirements Engineering (2004)
5. Salinesi, C., Etien, A., Wäyrynen, J.: Towards a Systematic Propagation of Evolution Requirements in IS Adaptation Projects. In: Australian Conference on Information Systems (ACIS), Hobart, Australia (December 2004)
6. Jackson, M.A.: System Development. Prentice-Hall, Englewood Cliffs (1983)
7. Jacobson, I., Booch, G., Rumbaugh, J.: The unified software development process. Eyrolles (2000)
8. Johson, B., Woolfolk, W.W., Miller, R.: Flexible Software Design. Auerbach Publications (2005)
9. Kroll, P., Royce, M.: Key Principles for Business-Driven Development. Rational Edge (2005)
10. Kroll, P., Kruchten, P.: The Rational Unified Process Made Easy: A Practitioner's Guide to the RUP. Addison Wesley, Reading (2003)
11. Lehman, M.M.: Laws of software evolution revisited. In: Montangero, C. (ed.) EWSPT 1996. LNCS, vol. 1149, pp. 108–124. Springer, Heidelberg (1996)
12. Lodding, K.N.: Hitchhiker's Guide to Biomorphic Software. QUEUE (June 2004)
13. Loucopoulos, P.: Engineering and Managing Software Requirements. Springer, Heidelberg (2005)

14. Meslati, D.: Mage: Une approche ontogénétique de l'évolution dans les systèmes logiciels critiques et embarqués. Phd thesis, University of Annaba (February 2006)
15. Meslati, D., et al.: The MAGE Ontogenetic Model: Towards autonomously-developed software. In: 17th Int. Conf. on Software & Systems Engineering and their Applications, Paris (November 2004)
16. Muller, P.A., Gaertner, N.: Modélisation objet avec UML. Eyrolles (2000)
17. Murch, R.: Autonomic Computing On Demand Series. Prentice Hall PTR, Englewood Cliffs (2004)
18. Sommerville, I.: Software Engineering. Edition (2000)
19. Rational Unified Process Fundamentals, Instructor Manual Version 2000.02.10, http://cjy.nyist.net

Embryonic Models for Self–healing Distributed Services[*]

Daniele Miorandi[1], David Lowe[1,2], and Lidia Yamamoto[3]

[1] CREATE-NET, v. alla Cascata 56/C, 38100 – Povo, Trento, IT
daniele.miorandi@create-net.org
[2] Centre for Real-Time Information Networks, University of Technology, Sydney
PO Box 123, Broadway 2007 NSW, Australia
david.lowe@uts.edu.au
[3] Computer Science Department, Bernoullistrasse 16, CH - 4056 Basel, Switzerland
Lidia.Yamamoto@unibas.ch

Abstract. A major research challenge in distributed systems is the design of services that incorporate robustness to events such as network changes and node faults. In this paper we describe an approach – which we refer to as *EmbryoWare* – that is inspired by cellular development and differentiation processes. The approach uses "artificial stem cells" in the form of totipotent nodes that differentiate into the different types needed to obtain the desired system–level behaviour. Each node has a genome that contains the full service specification, as well as rules for the differentiation process. We describe the system architecture and present simulation results that assess the overall performance and fault tolerance properties of the system in a decentralized network monitoring scenario.

Keywords: distributed services, autonomic computing, self–healing behaviour, robustness, embryogenesis, differentiation mechanisms.

1 Introduction

In this paper, we address the problem of devising architectures and methods for robust and self–healing distributed services. Given a service whose execution involves tasks running on a plurality of interconnected machines (or nodes), we introduce techniques for coping with faults and ensuring robustness at the system level.

The motivation for our work comes from the increasing utilisation of distributed services, i.e. services whose outcomes depend on the interaction of different components possibly running on different processors. Distributed services typically require complex design with regard to the distribution and coordination of the system components. They are also prone to errors related to possible

[*] This work has been partially supported by the European Commission within the framework of the BIONETS project EU-IST-FET-SAC-FP6-027748, www.bionets.eu

faults in one (or more) of the nodes where the components execute. This is particularly significant for applications that reside on open, uncontrolled, rapidly evolving and large–scale environments, where the resources used for providing the service may not be on dedicated servers (as the case in many grid or cloud computing applications) but rather utilise spare resources, such as those present in user's desktops or even mobile devices. (Examples of such scenarios are the various projects making use of the BOINC or similar platforms[1].) Other examples of distributed applications where each node takes on specific functionality include: peer-to-peer file sharing; distributed databases and network file systems; distributed simulation engines and multiplayer games; pervasive computing [13] and amorphous computing [1]. With all of these applications there is a clear need to employ mechanisms ensuring the system's ability to detect faults and recover automatically, restoring system–level functionalities in the shortest possible time.

In this paper we discuss an approach to addressing these issues that is inspired by cellular development and differentiation processes. Similar techniques have been applied in the evolvable hardware domain, giving rise to a specific research field called *embryonics* [10,11]. We propose an approach that utilises distributed nodes capable of differentiating into various types based on local knowledge, and which collectively lead to the emergence of the desired behaviour.

2   Background and Related Work

Robustness and reliability in distributed computing systems are well–studied topics. Classical fault–tolerance techniques include the use of redundancy (letting multiple nodes perform the same job) and/or the definition of a set of rules triggering a system reconfiguration after a fault has been detected [3]. When the scale of the system grows large, however, it is not practically feasible to pre–engineer in the system's blueprint all possible failure patterns and the consequent actions to be taken for restoring global functionalities. Such an issue is reminiscent of the reasons that led to the launch, by IBM, of the Autonomic Computing initiative [5], according to which one of the desirable properties of an autonomic system is *self–healing*. A self–healing system must recover full functionality, "healing" itself from faults and defects by actually fixing them autonomously, instead of just bypassing them.

In previous work by two of the authors [8,9], we considered the potential for using bottom-up approaches inspired by embryology to the automated creation and evolution of software. In these approaches, complexity emerges from interactions among simpler units. It was argued that this emergent behaviour can also inherently introduce self–healing as one of the constituent properties.

Our approach described in this paper builds on these concepts, leveraging off previous research conducted in the evolvable hardware domain on the application of architectures and methods inspired by the cellular developmental

---

[1] http://boinc.berkeley.edu/

and differentiation processes. Such approaches, which gave rise to the *embryonics* field [10,15], are based on the use of "artificial stem cells" [7,11], in the form of totipotent entities that can differentiate —upon reception of relevant signalling from nearby cells— into any component needed to obtain the desired system–level behaviour. Such an architecture has been successfully applied to field programmable gate arrays (FPGAs), resulting in the design of robust (i.e., able to sustain a large number of failures) and self–healing (i.e., able to recover automatically from faults by re-arranging its internal structures) hardware systems [10,14]. However, this earlier work did not consider the application of these approaches to distributed software applications.

In our work, we apply the concepts and tools developed in embryonics to the domain of *distributed software systems*. Such an application requires rethinking some of the design choices made by the embryonics research community to adapt to the specific features and constraints arising when working with software that is distributed over a network of interconnected machines. For example, network characteristics such as latency and dropped data packets can become a more significant factor in affecting the performance of the system. We call the resulting systems *EmbryoWare* (i.e. Embryonic Software).

Our approach bears many similarities to the work of Magrath [6], insofar as cells can propagate through a network and interact to achieve a global goal. The goal in Magrath's work is however to achieve a global pattern of cells (described as a phenotype) rather than a global behaviour. The cells in Magrath's approach also have a fixed set of behaviours that are not dependant upon the cell type (i.e. they do not change type). Further, their behaviour is dependant upon the network configuration of the neighbourhood rather than the cell types in the neighbourhood. This constrains the overall patterns of behaviour that can emerge from the network. The work described by Magrath does, however raise interesting questions with regard to how the cell genome can be designed so as to achieve a particular global pattern (or phenotype) – a question that is also relevant in our work.

Related approaches have also been recently proposed for enabling autonomic load–balancing among different application servers in a network [12]. Similar work by Chanprasert and Suzuki [2] considered the issue of self–healing in complex networks. Whilst not based on embryonics, the approaches were nevertheless bio-inspired (at the level of interacting individuals, rather than specialising cells), and demonstrated how distributed or decentralized control and processes of natural selection can lead to robust solutions.

3   Embryoware: Embryonic Software

EmbryoWare takes inspiration from the embryological or developmental processes in biology, by which an embryo made of initially identical cells (*stem cells*) develops into a full organism in which every cell assumes a different, specialized function, e.g. blood cells, skin cells, neurons. Stem cells are *totipotent*, i.e. unspecific and able to differentiate into the various cell types needed.

In EmbryoWare, *software stem cells* contain a genome with a concise representation of the *complete* service process to be performed. Such artificial stem cells are initially totipotent and are designed to spread throughout the network by self-replication. These cells differentiate into the various components needed for performing the overall service. Adequate signalling mechanisms shall be provisioned, so that cells could exchange information about the state of their neighbours. Upon detection of a fault in a neighbouring cell, they are able to re-enter the embryo state (unlike in biology), for differentiating again into the required functionalities, expressing the necessary genes.

It is important to note that, like most bio-inspired approaches, EmbryoWare remains an analogy, and is not meant to be entirely faithful to biology. A number of notable differences from true embryological processes can be highlighted. For example, unlike most biological examples, the EmbryoWare cells retain totipotency throughout their life, and are always able to re-differentiate into other cell types as needed. This is also done in other embryology-inspired approaches such as embryonics [10,15]. Similarly, real embryo formation makes use of on apoptosis (i.e. programmed cell death), whereas in our framework nodes will continue operation indefinitely (or until the operational task is completed).

3.1 Architecture and Components

The EmbryoWare approach is based on the use of nodes or cells (understood as basic computational units) [2] possessing the ability to decide autonomously, based on the task currently performed (referred to as *"cell type"* in the following) and on the ones performed by nearby cells, which task should be performed next in order to maximize the benefit for the system as a whole. Each cell is provided with a complete system–level specification of the service to be performed, called the *genome*, which includes:

(i)   a description of the expected behaviour of the service as a whole;
(ii) a description of the single tasks to be performed by cells;
(iii) a set of rules for deciding, based on the current task and the task performed by neighbouring cells, which task is to be performed next.

Each genome comprises a finite number of functions to be executed. We say that a cell performing a given function has *differentiated* into a given *type*. The type of a cell defines therefore its current role in the system–level architecture. A cell containing the genome can differentiate into any specific cell type encompassed by the genome.

Cells are arranged in a graph topology. The immediate neighbours (or 1-hop neighbors) of a cell are defined as the cells that are able to communicate directly with it. The n-hop neighbourhood consists of nodes located n communication hops away. In the embryonics domain, cells are often arranged in a toroidal regular grid, akin to a *cellular automaton (CA)*, where each cell is connected only

---

[2] Throughout the paper, we use the words *cell* and *node* interchangeably, as well as the words *task* and *function*.

with its immediate four (von Neumann neighborhood) or eight (Moore neighborhood) neighbours. Toroidal grids avoid border effects, but are rarely found in practical network scenarios (e.g. sensors spread over a field to be monitored) where border effects cannot be neglected. In our work we have not assumed this for the general case, and the algorithms are independant of the network topology.

An EmbryoWare system consists of the following components:

- *Genome*: defines the behaviour of the service as a whole, and determines the type to be expressed based on the local context (i.e., neighbouring cell types).
- *Sensing agent*: software component that periodically communicates with neighbours regarding their current type. The type of a cell is maintained in a separate register;
- *Replication agent*: software component that periodically polls the neighbours about the presence of a genome; if a genome is not present then the current genome is copied to the "empty" cell;
- *Differentiation agent*: software component that periodically decides, based on the cell's current type and the knowledge about the types of the neighbouring cells, which functions should be performed by the node.

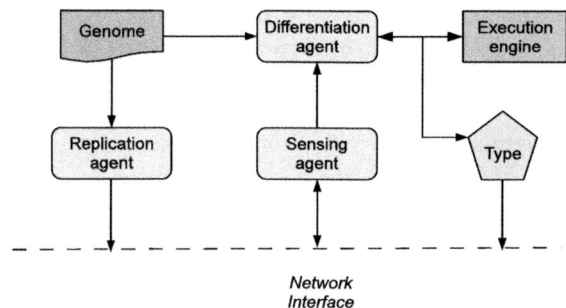

Fig. 1. Architecture of EmbryoWare: single–node view

A possible node–level architecture for EmbryoWare is shown in Fig. 1. The genome is connected to both the replication agent (which tries to replicate it in neighbouring "empty" cells) and to the differentiation agent. The latter also receives information from the sensing agent on the status of neighbouring cells. The cell's "type" is maintained in a separate register. It is also communicated to the execution engine, which performs the tasks/actions associated to the current type. The outcome of the execution process may trigger a differentiation (as in the case in which, e.g., the execution cannot be performed successfully due to some faults in the genome). The type of a cell can be read by the sensing agent of a neighboring node. The differentiation process can be implemented in a variety of ways. The simplest one is a lookup table (similar to those used in CAs), that determines, based on the current type of a node and on the sensed type of its neighbours, which type it should differentiate into.

In order to limit potential security issues, we further add *autonomy* constraints. Namely, a cell cannot be "reprogrammed" by a peer, but it will decide autonomously on the function to be performed (taking into proper account the local context). Hence a cell can influence only in an indirect way (by setting its own personal 'type') what neighbouring cells will do. Such a feature is appealing in that it limits the possibility of a malicious host affecting the emerging system–level behaviour.[3]

The two key aspects of EmbryoWare are related to (i) the development of an adequate representation of the service as a whole, able to be at the same time concise and expressive (ii) the development of efficient techniques for handling the differentiation process, requiring only local information from surrounding nodes. It is important to remark that the EmbryoWare architecture fits well services where the role of a given cell depends only on the type of neighbouring cells. While appropriate signalling mechanisms can be put in place to exchange information among remote nodes (enlarging in such a way the possible application domain), the resulting overhead may limit the system's performance. At the same time, communications among remote cells can be needed to obtain the desired system–level functionalities. Such a feature is supported by the system. What in EmbryoWare shall be limited is the signalling needed between cells in order to perform differentiation. We illustrate this issue in the case study presented later in the paper.

4   Algorithms for Embryonic Software

Given the architecture presented in the previous section, we may identify three key operations to be performed within an EmbryoWare–type system: *sensing*, *differentiation* and *replication*. In this section, we present algorithms for performing such functions in a distributed and asynchronous way.

4.1   Sensing Process

The sensing process is performed periodically at each node. Every τ_1 seconds, the cell issues a `queryType` message to its 1-hop neighbours, which reply sending information about their current type. The list of neighbours (indicated as `NeighboursList`) is created at bootstrap; its setup and maintenance is deferred to appropriate network–level services and is therefore not described in this work.

---

[3] It is however worth noticing that such an approach does not prevent malicious hosts from influencing the behaviour of the system. Proper security countermeasures should be put in place to limit the possible impact of such an occurrence. Further, it is worth remarking that the replication of genome in nearby cells require, in order to avoid potentially disruptive interference by malicious nodes, to put in place appropriate authorization and authentication procedures.

The information about the type of neighbours, Type($\cdot$), maintained in an appropriate knowledge base, is then updated. If a cell is faulty (i.e., machine is down due to maintenance or technical problems), it will not reply to the queryType message. Every node maintains therefore a timer, associated to a timeout, for each query message sent. In the case where no reply is obtained from a neighbour within the given timeout, its type is set to 'faulty'. The sensing process can be executed serially (polling one neighbour at a time) or in parallel (sending out queries to all neighbours and waiting for the message replies). Alg. 1 details the algorithm for the parallel case.

```
loop
    every τ₁
    for all i ∈ NeighbourList do
        send queryType message to i
        instantiate timer(i) for node i
    if timer(i) expires then
        Type(i) ← FAULTY
    if received message from i then
        update Type(i)
```

Algorithm 1. Sensing algorithm pseudo-code (parallel)

4.2 Differentiation Process

As with sensing, the differentiation process is performed periodically at each cell. We denote by τ_2 the differentiation period. Cells are not necessarily synchronized, so that the differentiation process can take place at different time instants at different nodes. In general, there is no need to specify a particular coupling between the differentiation period and the sensing period (they can well be implemented as independent threads). However, to reduce redundant processing, the period of the cell differentiation process should be equal to or larger than the sensing period τ_1.

The differentiation process is represented as a set of rules (which may be coded as a lookup table) provided as part of the genome.[4] Each cell uses information about its current type and the type of neighbouring cells to decide which type to express next (i.e., which function to be performed). The mechanism can be deterministic (given current state x and neighbours $1, \ldots, k$ in state $y_1, \ldots, y_k$, move to state z) or probabilistic (given current state x and neighbours $1, \ldots, k$ in state $y_1, \ldots, y_k$, move to state z_1 with probability p_1, to state z_2 with probability p_2 etc). A possible implementation of the differentiation algorithm is shown in Alg. 2.

---

[4] In general, other methods can be envisioned, based on, e.g., reaction–diffusion patterns [4]. It is also possible to envision accounting for environmental variables (such as, e.g., current CPU load or other contextual information) in the differentiation process. In this work, we limit our attention to a simpler set of rules only for the sake of simplicity.

```
loop
    every τ₂
    read Type(myID)
    for all i ∈ NeighbourList do
        read Type(i) {Update information on neighbour's type.}
    LOOKUP < Type(myID), Type(i₁), . . . , Type(iₖ)) >
    update Type(myID)
```

Algorithm 2. Differentiation algorithm pseudo-code

4.3 Replication Process

The replication process is meant to ensure that the system can make use of spare resources (empty cells that have not yet had a genome inserted into them by a neighbouring cell) whenever available, and hence the ability (at the system level) to recover from major faults. It could also be seen as a mechanism for automating service deployment in a distributed system. Through a suitable replication process, it would be sufficient to inject a "seed" genome into the system, and it will replicate itself across the network and differentiate into the necessary components. We assume that only the genome is replicated onto empty nodes: the management components (differentiation agent, replication agent, sensing agent, execution engine) are assumed to be present on all nodes in the system as part of the basic node platform. The replication algorithm works by periodically inquiring all neighbour cells about the presence of a genome. A node without an installed genome is still able to respond to queries, but will indicate that it has no functioning genome. If the cell is found to be empty, a copy of the genome is transmitted to the empty node, where it is installed and initiated. A possible implementation of the replication process is described in Alg. 3. While the replication period τ_3 is not strictly related to the sensing and differentiation period, the following relation provides an ordering suitable to maintain a good level of performance: $\tau_3 \gg \tau_2 \geq \tau_1$.

```
loop
    every τ₃
    for all i ∈ NeighbourList do
        send isGenomePresent message to i
        if noGenomePresent message received then
            send Genome to i
```

Algorithm 3. Replication algorithm pseudo-code

5 Evaluation on a Decentralized Monitoring Scenario

We evaluate the proposed techniques with a case study in the domain of decentralized network monitoring. Consider a sensor network, or any large network of devices where environment or system parameters must be monitored, and alarms must be raised whenever abnormal circumstances are detected. In such a scenario, nodes may perform different tasks, e.g. sense, collect, filter and log information, and then finally decide whether an alarm should be raised or not, based on the information sensed. This is a typical scenario where the automatic differentiation into each of these separate tasks, performed by EmbryoWare, is a helpful feature in order to keep providing a prompt and reliable service in spite of node failures or unexpected network changes.

5.1   Case Study Description

We consider a network of cells performing resource monitoring, logging, and alarm generation. We assume that each cell possesses some parameter that needs to be periodically monitored. Each *monitoring* sample needs to be reported to a *logging* cell that should be no more than 2 hops away from the *monitoring* cell in order minimise network traffic[5]. The *logging* cell will accumulate data samples and report them periodically to *alarm* cells. The network should contain 2 *alarm* cells for redundancy, but no more than 2 *alarm* cells in order to minimise the resource requirements associated with alarms. Each *alarm* cell should have two 1–hop neighbours that are *analysis* cells which it uses to assist in analysing the provided samples. When the alarm cell recognises an alarm condition on one of the data samples this is reported in a system-dependant fashion.

In addition, in order to distribute the load associated with alarm condition evaluation, the cells taking on the *alarm* behaviour should change periodically. Further, to evaluate fault performance, our simulation includes random genome failures (with a predefined probability that is independent in our case study of the cell type). A genome fault is where the genome can no longer operate correctly – and hence will not respond with a valid genome type when queried by a neighbouring node. The underlying management components are however still operational, and so a neighbour could reinsert a new genome to correct the genome fault. Conversely, a node failure is where the node becomes permanently inoperable.

5.2   Cell Types and Their Behavior

The desired behaviour requires a genome with the following types: Stem, Faulty, Monitor, Logger, Analysis, Alarm. A *stem* cell in our system is one that is currently idle but ready to differentiate into some needed type. The specific behaviours outlined above can then be obtained in a number of different ways, with different implications for the cell type patterns that emerge, and the timing within which the differentiation happens.

A crucial aspect is how to maintain a global target number of alarm nodes ($N = 2$ in the case study), in a decentralized way. Presently, this is achieved by letting nodes broadcast a beacon when they become alarm ones. Each cell maintains a list of active alarm nodes. When any logging cell is ready to send its data, it will try to send to each alarm cell that is in its list, and if it does not receive an acknowledgment then it removes that alarm from its list.

---

[5] A 2-hop neighbourhood can be monitored by asking 1-hop neighbours about their neighbours. i.e. when a 1-hop neighbour reports its node type, in also includes relevant information about it's own 1-hop neighbours.

One key choice is with regard to the processes associated with differentiation. It is possible to implement the conversion of a cell's genome into an *alarm* type either proactively or reactively. In the pro-active case, each cell (through its sensing agent) will monitor the state of other cells and if it determines that there are not two *alarm* cells, then it can proactively differentiate into an *alarm* cell (albeit with a level of randomness to ensure that not all nodes differentiate into an *alarm* cell simultaneously). In the reactive case, when a *logging* node's execution engine attempts to transmit data to the *alarm* cells, if it receives no response then it can reactively trigger the differentiation into an *alarm* cell. Both variations have been implemented in order to compare the performance of the two approaches. The detailed implementation description is deferred to App. A.

5.3   Results and Assessment

The case study has been implemented and simulated with Matlab, using a regular Moore-neighbourhood grid. It illustrates the viability of both the general approach and the specific architecture that has been proposed. It also uncovers a number of performance considerations and system design guidelines.

As a first performance metric, we considered the fraction of time the system was in the 'up' state (i.e., a state in which the requirements in terms of presence of *loggers*, *alarms* and *analysis* nodes were met) as a function of the failure rate of single genomes. We considered a 10×10 network, *alarms* differentiating back to *stem* cells after 20 s, *stem* cells becoming *monitors* with probability (per second) of 0.1, *loggers* differentiating back to *stem* cells with probability (per second) of 0.001, *alarms* differentiating to *stem* cells with probability (per second) of 0.1, *alarms* decaying to *stem* cells with probability of 0.3 in case too many *alarms* are present in the system. At the beginning of the simulation, one single genome is injected into the node at position $(1, 2)$, and is left to replicate and differentiate in the system. For each value of the genome fault rate, 20 runs were performed, each one consisting of 3000 iterations of the differentiation process over the whole system. The genome fault rate (in s^{-1}) was varied between 10^{-3} to 0.9. The results obtained are plotted in Fig. 2 on a semi–logarithmic scale. As it can be seen, the fraction of time spent in an invalid state is only marginally sensitive to the genome fault probability, an appealing feature for system's designers. The presence of a "floor" on the system downtime depends on the high level of randomness present in the system; in particular the decay of nodes into the *stem* cell state leads to a non–zero probability of being in an invalid state even in the absence of genome faults.

As a second performance metric, we considered the time necessary for the system to reach a valid state (i.e., able to meet the requirements in terms of presence of *loggers*, *alarms* and *analysis* nodes) starting from a single genome injected into the cell $(2, 1)$. Such a parameters provide a measure of the time needed for a newly deployed system to settle into a valid state. We varied the network size between 36 and 400 nodes and used a genome failure rate of 0.01 s^{-1}. For each network size considered, 10 independent runs were executed. The other parameters were set as described above. The results obtained are plotted in

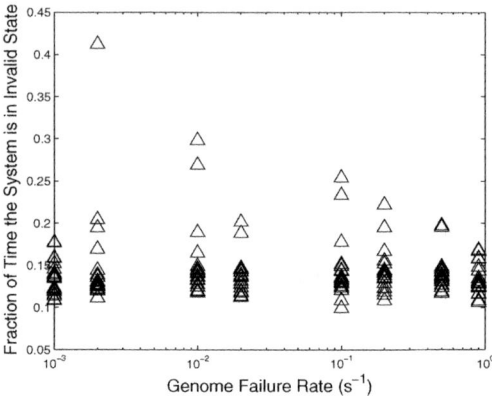

Fig. 2. Fraction of time the system is in an invalid configuration as a function of the genome fault rate

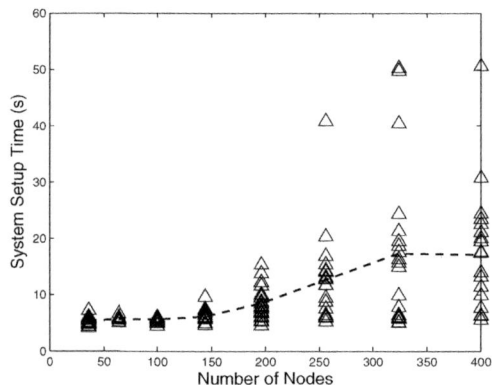

Fig. 3. Time necessary for reaching a valid configuration starting with a single genome injected in node $(2, 1)$ as a function of the network size

Fig. 3. The dashed line represents the mean values, while the outcomes of single runs are reported using triangular markers. As it can be seen, the time to reach a valid state increase with the network size, growing from ~ 5 s to ~ 15 s (in terms of mean values). Such an increase cannot be ascribed to communication delays, but is related to the fact that the probabilistic differentiation processes at the hearth of the example shown require careful tuning of parameters to offer good performance for different network size. In other words, the parameters driving the transitions between different types should be tuned according to the network size in order to achieve optimal performance. At the same time, while an "aggressive" behaviour (with rather high values for differentiation probability, and hence nodes volunteering more rapidly) can lead to a speed–up of the time taken to reach a valid state, it may also lead to undesirable oscillation during normal operations. This is because too many nodes can differentiate into a particular role (e.g. the alarm in our scenario). When each node realises that there are too many volunteers, they then "aggressively" revert back - though again too many do so, setting up a cycle. This is a combination of high probabilities of conversion, coupled with the communication delays, meaning that cells make a differentiation decision before they have data on the fact that they have neighbours who have also done so. We are yet to investigate the circumstances under which such an oscillatory beahviour may occur.

Overall, the simulation demonstrated that robust fault tolerance, in the event of both node and genome faults, can be supported quite elegantly. In the case of genome faults, provided the node can detect a fault in its genome, it can purge it and allow the replication process to reinsert a new (operational) one. When a node itself fails then it will remain inoperable, but the simulation demonstrated that its functionality is subsequently accommodated by other ones.

The simulation also highlighted that in the current architecture the optimal tuning of the differentiation parameters is dependant upon the network size. For example, when an *alarm* differentiates back to a stem cell, *logger* nodes will probabilistically differentiate to form new *alarm* nodes. As the network size increases, the number of differentiations that occur, for the same probability parameters, will increase. This behavioural dependence upon network size is undesired – ideally the performance should be independent of the number of nodes. [6]

A number of genome design issues also emerged from the simulation. It became evident that network fringe effects need to be considered in designing the genome behaviours. For example, with a poorly designed genome, it is possible that an *alarm* node might appear on the border of the network where the requirement for two neighbouring *analysis* nodes cannot be met. This issue could be addressed in several ways.Nevertheless, this does highlight that careful consideration must be given to the genome design lest unintended behaviours emerge.

Another interesting behaviour that became apparent was what we might call the *Hydra*[7] behaviour. If we split the network of nodes into two isolated subnetworks, then the nodes differentiate in order to create a fully operational system in each sub-network, exhibiting a natural self-healing ability.

6   Conclusion

In this paper we have proposed *EmbryoWare*, an embryonic–inspired architecture for robust and self–healing distributed software. The approach is based on

---

[6] Such a dependence on the network size comes from the fact that the case study has to satisfy a *global* constraint (i.e., on the number of *alarm* nodes in the system). If only *local* constrains were present (e.g., one alarm cell shall be present within k hops from any loggers), the dependence on the network size would blur.

[7] The Hydra is a small freshwater animal that exhibits an interesting behaviour. If it is severed into multiple parts, then each part is capable of morphollaxis – i.e. reorganising / regenerating to become a fully functioning individual hydra.

each node in the system containing a genome that includes a complete specification of the service to be performed, as well as a set of rules that ensure each node differentiates into the node type required to provide required overall system behaviour. The simulations that we have performed have examined the case of genome failure and demonstrated the viability of this approach as well as the inherent robustness and self-healing that is achieved.

In ongoing work we will be considering other failure scenarios (e.g. link and node failures, changes in network topology, etc.) and how the system recovers from these failures. We will also be broadening the basis for analysing the system performance to include a more thorough analysis of how the tuning of the genome differentiation rules – and especially the stochastic parameters associated with decisions on the timing of the differentiation – affect the overall performance. We will also be evaluating the processing and communication overheads that this approach introduces, and how these scale with changes in the network size.

A number of additional research questions also emerge from these preliminary studies. Whilst our simulation captured relatively sophisticated behaviour, it was still less complex than many applications. It does raise the question of how complex a genome needs to be in order to provide desired behaviours, and whether a threshold will be reached where the genome complexity becomes prohibitive. Our evaluation also indicated the importance of considering carefully the processes required to understand and design for reliability and robustness – particularly in the context of network fringe effects. Subsequent work will also need to consider how to handle differences in the capabilities of nodes by adequately taking them into account in the differentiation process. The system sensitivity to various environmental characteristics, such as network size and network latency should also be considered.

Other future work could be to complement the EmbryoWare framework with an apoptosis or programmed cell death scheme, as a reverse operation for the current replication scheme. Apoptosis could be useful to optimize the placement of redundant functions (e.g. to minimize broadcast, etc.). It could also be used to deal with security breaches by isolating and killing misbehaving cells, a mechanism that is necessary when code can propagate in the network by replication. Apoptosis could also offer a mechanism by which, once the processing is complete in some regions of the network, nodes could "die" elegantly. Finally, another topic for future work would be to actually evolve the genome program to adapt to new situations.

References

1. Abelson, H., Allen, D., Coore, D., Hanson, C., Homsy Jr., G., Knight, T.F., Nagpal, R., Rauch, E., Sussman, G.J., Weiss, R.: Amorphous computing. Communications of the ACM 43(5), 74–82 (2000)
2. Champrasert, P., Suzuki, J.: A biologically-inspired autonomic architecture for self-healing data centers. In: 30th IEEE International Conference on Computer Software and Applications Conference (COMPSAC), vol. 2, pp. 350–352. IEEE, Los Alamitos (2006)

3. Coulouris, G.F., Dollimore, J., Kindberg, T.: Distributed systems: concepts and design. Addison-Wesley Longman, Amsterdam (2005)
4. Deutsch, A., Dormann, S.: Cellular automaton modeling of biological pattern formation: characterization, applications, and analysis. Birkhäuser, Basel (2005)
5. Kephart, J.O., Chess, D.M.: The vision of autonomic computing. IEEE Comp. Mag. 36(1), 41–50 (2003)
6. Magrath, S.: Morphogenic systems engineering for self-configuring networks, July 2-5 (2007)
7. Mange, D., Stauffer, A., Tempesti, G.: Embryonics: a microscopic view of the molecular architecture. In: Sipper, M., Mange, D., Pérez-Uribe, A. (eds.) ICES 1998. LNCS, vol. 1478, pp. 185–195. Springer, Heidelberg (1998)
8. Miorandi, D., Yamamoto, L.: Evolutionary and embryogenic approaches to autonomic systems. In: Proc. of ValueTools (InterPerf Workshop), Athens, Greece, pp. 1–12 (2008)
9. Miorandi, D., Yamamoto, L., De Pellegrini, F.: A survey of evolutionary and embryogenic approaches to autonomic networking. Computer Networks (in press, 2009), doi:10.1016/j.comnet.2009.08.021
10. Ortega-Sanchez, C., Mange, D., Smith, S., Tyrrell, A.: Embryonics: a bio-inspired cellular architecture with fault-tolerant properties. Genetic Programming and Evolvable Machines 1, 187–215 (2000)
11. Prodan, L., Tempesti, G., Mange, D., Stauffer, A.: Embryonics: artificial stem cells. In: Proc. of ALife VIII, pp. 101–105 (2002)
12. Saffre, F., Shackleton, M.: "embryo": an autonomic co-operative service management framework. In: Artificial Life XI: Proc. 11th Int. Conf. Simulation and Synthesis of Living Systems, pp. 513–520. MIT Press, Cambridge (2008)
13. Saha, D., Mukherjee, A.: Pervasive computing: A paradigm for the 21st century. Computer 36(3), 25–31 (2003)
14. Stauffer, A., Mange, D., Tempesti, G., Teuscher, C.: A Self-Repairing and Self-Healing Electronic Watch: The BioWatch. In: Evolvable Systems: From Biology to Hardware. LNCS, vol. 2210, pp. 112–127. Springer, Heidelberg (2001)
15. Tempesti, G., Mange, D., Stauffer, A.: Bio-inspired computing architectures: the *embryionics* approach. In: Proc. of IEEE CAMP (2005)

A   Detailed Use–Case Implementation Description

To illustrate a specific Genome pattern, we describe the execution and differentiation behaviours for the case where the genome responds reactively to failures to find *logger* and *alarm* nodes. Algorithm 4 describes the execution behaviours. As can be seen, a *monitor* node can reactively trigger a differentiation into a *logger* node when required, and a *logger* node can reactively trigger a differentiation into an *alarm* node when required. Algorithm 5 shows both the proactive and reactive differentiation behaviours for the genome. The proactive differentiation is triggered by the relevant sensing of the node neighbourhood, whereas the reactive differentiation is triggered by events occurring in the execution engine.

- *Stem* cell:
    - None
- *Faulty* cell:
    - None
- *Monitor* cell:
    - every T_M generate sample S
    - **repeat**
        - transmit S to *logger*
        - **if** transmission failed **then**
            - search 2-hop neighbourhood for logger
            - **if** no *logger* found **then**
                - trigger reactive differentiation
    - **until** S processed
- *Logger* cell:
    - accept, record all *Monitor* samples
    - accept, register all *Alarm* beacons
    - every T_L
    - **while** recorded samples still to be transmitted **do**
        - attempt transmit samples to all *alarms*
        - **for all** *alarm* cells that do not respond **do**
            - deregister *alarm*
        - **if** < 2 registered *alarms* **then**
            - trigger reactive differentiation.
- *Analysis* cell:
    - accept, process *Alarm* requests
- *Alarm* cell:
    - accept, process *Logged* data
    - send *Alarm* requests to analysis cells

Algorithm 4. Execution behaviours for typical Genome for data logging application.

- *Stem* cell:
    - Proactive: with probability $P_{TtoM} \Rightarrow Type \leftarrow Monitor$
- *Faulty* cell:
    - None
- *Monitor* cell:
    - Reactive: no logger $\Rightarrow Type \leftarrow Logger$
    - Proactive: *alarm* neighbour has < 2 *analysis* cells $\wedge$ offer accepted $\Rightarrow Type \leftarrow analysis$
    - Proactive: probability $P_{MtoT} \Rightarrow Type \leftarrow Stem$
- *Logger* cell:
    - Reactive: if < 2 alarms found, with probability $P_{LtoA} \Rightarrow Type \leftarrow Alarm$, broadcast beacon
    - Proactive: with probability $P_{LtoT} \Rightarrow Type \leftarrow Stem$
- *Analysis* cell:
    - Proactive: *alarm* not responding $\Rightarrow Type \leftarrow Stem$
- *Alarm* cell:
    - Proactive: > 2 registered *alarms* $\wedge$ probability $P_{MAtoT} \Rightarrow Type \leftarrow Stem$
    - Proactive: active for $> T_{Alm} \wedge$ probability $P_{MtoT} \Rightarrow Type \leftarrow Stem$

Algorithm 5. Differentiation behaviours for typical Genome for data logging application.

Bio-inspired Speed Detection and Discrimination

Mauricio Cerda[1], Lucas Terissi[2], and Bernard Girau[1]

[1] Loria - INRIA Nancy Grand Est, Cortex Team
Vandoeuvre-lès-Nancy - France
{cerdavim,girau}@loria.fr
[2] Laboratory for System Dynamics and Signal Processing,
Universidad Nacional de Rosario - CIFASIS - CONICET - Argentina
terissi@cifasis-conicet.gov.ar

Abstract. In the field of computer vision, a crucial task is the detection of motion (also called optical flow extraction). This operation allows analysis such as 3D reconstruction, feature tracking, time-to-collision and novelty detection among others. Most of the optical flow extraction techniques work within a finite range of speeds. Usually, the range of detection is extended towards higher speeds by combining some multi-scale information in a serial architecture. This serial multi-scale approach suffers from the problem of error propagation related to the number of scales used in the algorithm. On the other hand, biological experiments show that human motion perception seems to follow a parallel multi-scale scheme. In this work we present a bio-inspired parallel architecture to perform detection of motion, providing a wide range of operation and avoiding error propagation associated with the serial architecture. To test our algorithm, we perform relative error comparisons between both classical and proposed techniques, showing that the parallel architecture is able to achieve motion detection with results similar to the serial approach.

Keywords: motion perception, optical flow, speed discrimination, MT.

1 Introduction

The visual capabilities in humans have motivated a large number of scientific studies. The performance to perceive and interpret visual stimuli in different species including humans is outstanding: the wide range of tolerance to different illumination and noise levels are just a few characteristics that we are aware of, but that are still barely understood.

An important aspect in visual processing is the perception of motion. Motion is a key step in several computer vision tasks such as 3D reconstruction, feature tracking, time-to-collision estimation, novelty detection, among others [1]. Motion is also one of the features that many species can perceive from the flow of visual information, and its detection has been observed in a large number of animals [2], from invertebrates to highly evolved mammals. From optical

E. Altman et al. (Eds.): Bionetics 2009, LNICST 39, pp. 167–176, 2010.
© Institute for Computer Sciences, Social-Informatics and Telecommunications Engineering 2010

engineering and experimental psychology we already know the main features of human motion discrimination [3]. In this work, we are particularly interested in taking inspiration from biology in order to design a parallel algorithm with similar discrimination capabilities as obtained by classical serial architectures.

Our work begins by presenting an overview of techniques to detect motion in machine vision to continue with the available experimental results and their procedures in human psychophysics. Section 3, presents our algorithm for speed detection with which we perform our simulations, and compare with experimental data. The results are analyzed in section 4, and in the last two sections, we present the discussion and conclusions about our work.

2   Overview

In this work we are interested in the detection of motion, specifically in the coding and retrieval of speed (v), and in the link between the idea of selecting a range of speed to work with, and providing wider ranges of discrimination as observed in human psychophysics experiments [4]. We focus on two features: the multi-scale architecture of the speed detection, and the relation between the number of multi-scale levels and the range of speeds the system is sensitive to.

2.1   Motion Detection in Computer Vision

The detection of motion is a widely used operation in computer vision. Commonly called "optical flow extraction", the main objective is to assign a vector $v = (u, v)$ to each frame pixel from a given sequence of frames (at least two). In this section, we explain the basic technique to increase the motion range of an optical flow extraction which the method is sensitive to. We ground our explanation on the well-known Lucas & Kanade's method [5,1] (the basic multi-scale technique similarly applies to other methods for optical flow extraction).

Optical flow. Many optical flow extraction methods are based on the initial assumption of brightness conservation, that is,

$$\frac{dI(x, y, t)}{dt} = \frac{\partial I}{\partial x} u + \frac{\partial I}{\partial y} v + \frac{\partial I}{\partial t} = 0 \tag{1}$$

where $v = (u, v)$ is the velocity vector. A well known technique following this approach is the Lucas & Kanade algorithm [5], that minimizes the following cost function in a small fixed region Ω, *i.e.*

$$v = \arg\min_{v} \left\{ \sum_{x \in \Omega} W^2(x) \left[\nabla I(x, t) \cdot v + \frac{\partial I}{\partial t}(x, t) \right]^2 \right\} \tag{2}$$

where W is a two-dimensional Gaussian function used to give more importance to the central points and Ω is a square region of a few pixels. This minimization estimates v with sub-pixel precision after a few iterations. This method achieves good optical flow extraction in regions where $|\nabla I(x, t)| > 0$, such as corners [6].

Serial multi-scale optical flow. The Lucas & Kanade method for optical flow extraction considers a small region Ω. The use of this region Ω is not particular to this method: it is used in most algorithms [1]. As the computation is performed in small windows, the detection of motion is constrained to detect speeds up to ω pixels per frame, where ω stand for the diameter of Ω. To overcome this limitation, a multi-scale representation of the images can be performed, usually by considering Gaussian pyramids [7]. A Gaussian pyramid representation of an image is computed by recursively smoothing (using a Gaussian kernel) and sub-sampling the original image. In this way, the original image is represented by a set of smaller images. The representation at scale level $l = 0$ is the original image itself. The image at level l is obtained by sub-sampling a filtered version of the image at level $l - 1$ with a down-sampling factor equal to 2. Thus, the size of the image at each level l is $N_l = N_{l-1}/2$ with $l = 1, 2, \ldots, (L - 1)$, where L is the number of levels of the representation.

In the serial multi-scale optical flow estimation, speed is computed by sequentially projecting the estimation obtained at level l to level $l - 1$, until level $l = 0$. There are complex strategies for computing the optical flow with a multi-scale approach [6]. A simple solution for optical flow computation is implemented in the widely used computer vision library OpenCV [8,7]. In this case, the multi-scale estimation starts from the highest level ($l = L - 1$) and it propagates to the next one:

$$v_{l-1} = 2 * v_l + d_{l-1}(v_l) \qquad (3)$$

where d_{l-1} is the estimation of velocity at level $l - 1$ after projecting the estimation v_l by warping the image by $-v_l$ at level $l - 1$. Computing the optical flow from the highest level and then projecting the solution to the lower level [6,7] increases the range of detectable speeds. This range is wider when more scales are used. On the other hand, the sequential projection between levels also propagates the error introduced at each level. Thus, in terms of precision, increasing the number of scales in the representation increases the error introduced in the estimation.

2.2   Biological Elements

This section sketches out the current experimental knowledge in biology, focusing on studies of speed coding in the human brain [9] and on higher level descriptions of speed discrimination from experimental psychophysics [3,10,11].

Parallel architecture. In the human brain the main area that is responsible for coding different speeds is area MT [2]. It is located in the occipital region (back of the head). Neurons in this area are selective to stimuli moving at a given speed [12]. Their spatial organization is retinotopical [13]: each neuron has a reduced visual field, and neurons who share the same local visual field are grouped together in some macro-column that contains cortical columns that are selective for different orientations. This configuration allows a complete mapping of the visual field with a group of cortical columns that codes for all possible directions of local motions. The spatial organization is less known with respect to the speed selectivity. Nevertheless, it has been found that (1) the average detected speed increases with eccentricity (with respect to the retinotopical organization of MT), (2) similar speeds are detected by neurons closer than for distinct speeds, and (3) for each eccentricity, there are neurons for different speeds [13]. The interactions between different units is not completely understood, but there is evidence that units sensitive to different speeds could be coding a range of speeds in parallel [9]. It has been observed that the range of detectable speeds is not uniformly covered [12], but in this work we are interested in the simultaneous existence of speed selective units in MT that could be accountable for a parallel architecture dealing with different speeds.

Speed discrimination. In the work of McKee et al. [3], two subjects were exposed to several stimuli, one of those being a horizontal scaled single bar vertically moving at different eccentricities. The goal of this experiment was to determine the minimal relative detectable variation in speed for every subject with the sight fixed at a certain location and for each stimuli eccentricity.

It is important to mention how this was actually measured, because the subject cannot assign a precise velocity at each location. Instead, given a reference velocity, the subject was asked to indicate whether the next presented stimuli moves faster or slower. The minimal detectable variation was then statistically inferred. Related experiments were performed by others [4,10,11], showing that the measurements are not affected by different contrast conditions, and that they do not depend on binocular or monocular sight.

The described experiments study the speed discrimination at several eccentricities[1], see Fig. 1. In this work we are interested in each one of these eccentricities and their related discrimination properties, and not in the relations between

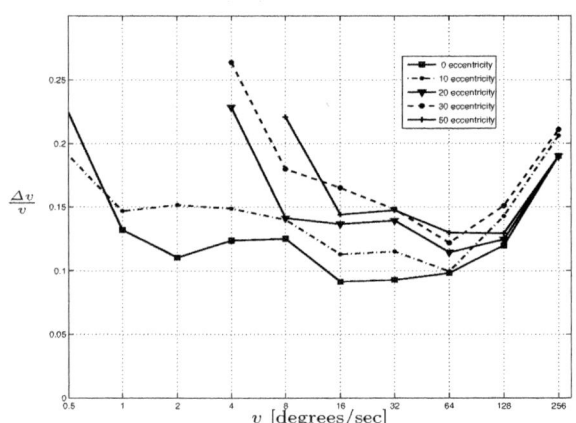

Fig. 1. Weber fraction (minimum threshold of perceived change in speed or $\Delta v/v$) as function of velocity, for different eccentricities in the human subject B.D.B. (condition 2), in the differential motion experiment by [4] (reproduced from their work), each curve is at eccentricities 0,10,20,30 and 50, respectively (left to right). The speed axe is in logarithmic scale.

---

[1] Distance to the center of the eye in foveated vision (humans, primates and others).

different eccentricities. In order to model these discrimination functions, we need to generate a given discrimination percentage in a range of speed $[v_1, v_2]$. We also point out that the left side of the experimental curves, see Fig. 1, is related to the eccentricity but the same idea holds: for each eccentricity there is a wide range of speed discrimination, where the relative error (rather than the absolute error) remains stable (5%-15%).

3 Proposed Parallel Multi-scale Speed Detection

Multi-scale speed detection is based on the fact that a particular speed detection algorithm can be used to estimate slower speeds at lower levels and to estimate faster speeds at higher levels. This information is used in the above described serial multi-scale optical flow algorithm to detect speeds in a wide range of velocities by projecting the information at level $l+1$ to estimate speed at level l, i.e. in a serial manner. As it is described in [12,9], it seems that human motion perception is based on a parallel multi-scale scheme. Based on this idea, the speed detection algorithm proposed in this paper estimates the speed by combining the information computed at each level independently, i.e. using the multi-scale information in a parallel manner. In this case, there is no error propagation on the computation of speeds at each level because it does not depend on the estimation performed for other levels. At each level l, we compute speeds using the optical flow estimation algorithm described above, see subsection 2.1. As explained before, this choice does not bias our results, since our work is to provide a bio-inspired parallel speed detection instead of the standard serial approach, for any optical flow extraction method.

As expected, the speed detection algorithm estimates speed with a certain error at each multi-scale level l. The confidence in the estimation of speed v_r at level l, denoted as $k_l(v_r)$, can be defined as

$$k_l(v_r) = 1 - \left| \frac{v_r - v_e}{v_r} \right| \qquad (4)$$

where v_r is the magnitude of the object's real speed ($v_r = \|\boldsymbol{v}_r\|$) and v_e is the magnitude of the average estimated speed on the object pixels location. It can be noted, that this computation only takes into account the magnitude of the speed, ignoring its direction. Figure 2(a) shows the confidence $k_l(v_r)$ for three different multi-scale levels. These distributions were computed using an input image sequence containing an object moving at different speeds in a range from 0.5 pixels per frame to 20 pixels per frame. To statistically determine the confidence at each level l and speed v_r, the experiments were carried out using the input image sequence with several realizations of Gaussian white noise, then the resulting confidence $k_l(v_r)$ is computed as the mean value of the ones obtained in the experiments. Figure 3(a) shows two frames of an input image sequence used in the experiments. In this sequence the object is moving at 10 frames per pixel in the bottom-right direction.

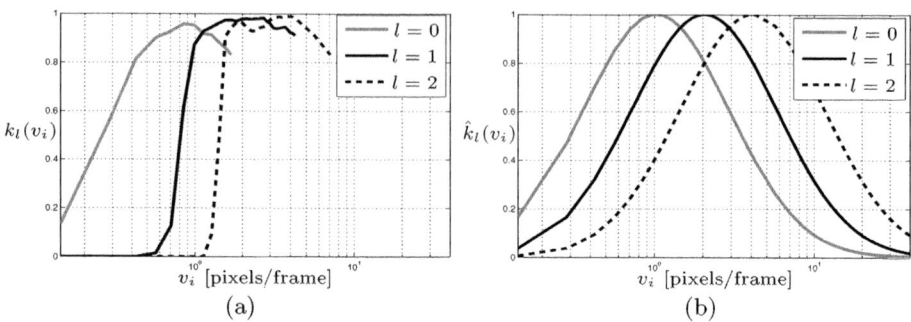

Fig. 2. Confidence distribution k_l for different levels l. (a) Experimental distributions k_l. (b) Approximated distributions $\hat{k}_l$.

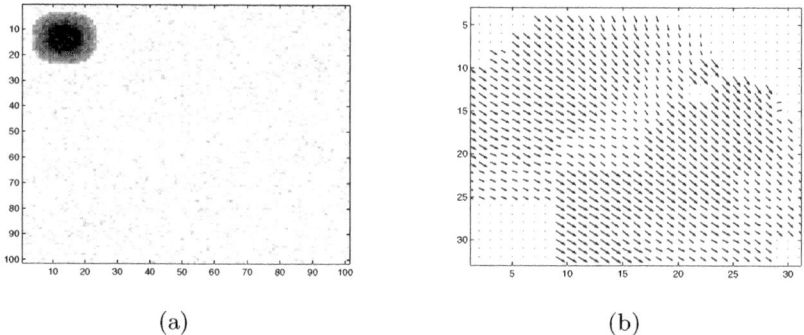

Fig. 3. In (a) one frame of an input image sequence used in the experiments is depicted ($v = 10$ bottom-right direction). (b) is the obtained optical flow using the proposed parallel multi-scale algorithm. Note that optical flow image was zoomed around the object position.

As it may be seen in Figure 2(a), a particular speed v_r can be detected at several multi-scale levels but with different confidence values. Thus, the current speed could be estimated by taking into account the speeds computed at each level l and their associated confidence values k_l. For that reason, the experimental distributions depicted in Fig. 2(a) have to be approximated by a closed-form equation. In this work, these distributions are approximated (modeled) as Gaussian distributions in a semi-log space defined by the following equation

$$\hat{k}_l(v_r) = \exp\left(-\left[\frac{\log(v_r) - \mu_l}{\sigma_l}\right]^2\right) \qquad (5)$$

$$\mu_l = \mu_0 + \log(c^{l-1}) \qquad (6)$$

$$\sigma_l = \sigma_0 \qquad (7)$$

where μ_0 and σ_0 are the mean and variance of the distribution at level $l = 0$ and c is the scaling factor used in the sub-sampling of the images. The approximated distributions $\hat{k}_l$ for $l = 0$, $l = 1$ and $l = 2$ are depicted in Fig. 2(b). It may be seen that a better approximation of the distributions could be obtained using a particular set of variables for each level but this would increase the model complexity. The approximation of the distributions $\hat{k}_l$ for each level in Eq. (5) only depends on μ_0, σ_0 and c. It may be noted that this approximation allows to perform the estimation of speeds using different values of the scaling factor c, which is usually set to $c = 2$, *i.e.*, the case of using Gaussian pyramids for the sub-sampling.

Finally, denoting the detected speed at each level l by $\boldsymbol{v}_e^l$, the proposed algorithm computes the current speed, using the speed detected at each multi-scale level with its associated confidence value $\hat{k}_l(\|\boldsymbol{v}_e^l\|)$, as

$$\boldsymbol{v}_f = \frac{\sum_{l=0}^{L-1} \boldsymbol{v}_e^l \, \hat{k}_l(\|\boldsymbol{v}_e^l\|)}{\sum_{l=0}^{L-1} \hat{k}_l(\|\boldsymbol{v}_e^l\|)} \tag{8}$$

where L is the number of levels used to compute the estimated speed $\boldsymbol{v}_f$. Figure 3(b) shows the obtained optical flow using the proposed parallel multi-scale algorithm. The comparison between the experimental confidence distribution of the proposed algorithm and confidence distributions for three levels is shown in Fig. 4. As expected, the confidence distribution of the parallel multi-scale algorithm with $L = 3$ is approximately the envelope of the confidence distributions of levels $l = 0$, $l = 1$ and $l = 2$.

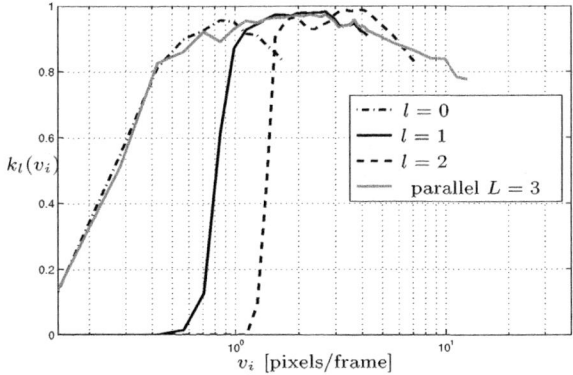

Fig. 4. Comparison between confidence distribution of the proposed algorithm (parallel) with $L = 3$ and $c = 2$, and confidence distributions k_l for levels $l = 0, 1$ and 2 without projection between them

4   Results

As it was described in subsection 2.2, speed discrimination is computed as the minimal detectable variation in speed of a particular visual stimuli. In this work,

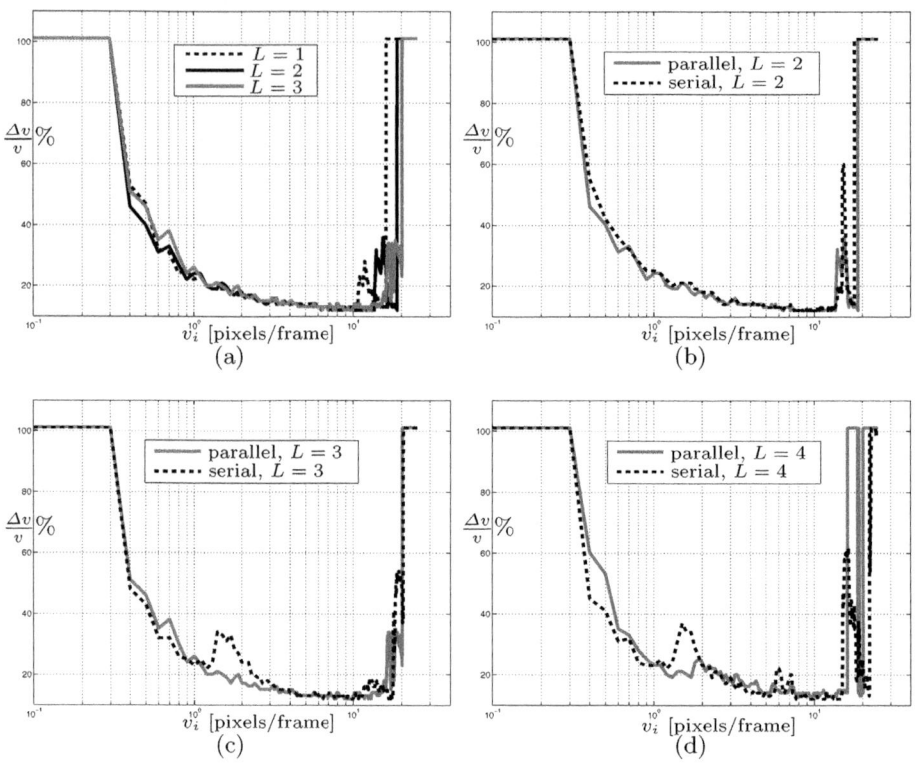

Fig. 5. (a) Discrimination of the proposed algorithm for $L = 1$, $L = 2$ and $L = 3$ ($c = 2$). In (b), (c) and (d), the comparisons between the discrimination of the proposed parallel ($c = 2$) and the serial algorithms for $L = 2$, $L = 3$ and $L = 4$, respectively, are shown

a variation in speed, from a given reference speed v_{obj}, of the moving object is considered to be noticeable if the following inequality holds

$$\left| \frac{\widehat{v}_{v_{obj}} - \widehat{v}_{v_{obj} \pm \Delta v_{obj}}}{v_{obj}} \right| > \alpha \tag{9}$$

where $\widehat{v}_{v_{obj}}$ and $\widehat{v}_{v_{obj} \pm \Delta v_{obj}}$ are the speeds estimated by the algorithm when the object is moving at velocities v_{obj} and $v_{obj} \pm \Delta v_{obj}$, respectively, and α is the percentage of variation from the object speed v_{obj} required to consider Δv_{obj} as detectable. It may be noted in Eq. (9) that a variation in speed is considered noticeable if it is detectable when v_{obj} is both increased and decreased in Δv_{obj}. To statistically determine the minimum value of Δv_{obj} several experiments were carried out using the input image sequence with several realizations of Gaussian white noise. Then, the minimal detectable variation in speed, from a given reference speed v_{obj}, is computed as the minimum detectable Δv_{obj} obtained in 90% of the experiments.

We summarize our results in Fig. 5. Figure 5(a) shows the discrimination of the proposed parallel multi-scale algorithm for different values of L. The range of discriminated speeds is enlarged when the number of levels used in the multi-scale representation increases. In comparison with the serial multi-scale, our method has a similar range of speed discrimination when the same number of levels are used, see Fig. 5(b), 5(c) and 5(d). Considering both mean and variance of the discrimination in the range of speeds from 1 to 15 pixels per frame, the parallel multi-scale method shows lower values. For the case of $L = 3$, parallel discrimination has mean= 14.1 and variance= 2.2, while serial discrimination has mean= 15.5 and variance= 4.2. This indicates that the proposed parallel algorithm presents a better discrimination in this range.

5 Discussion

Recent works [6,14] have developed the idea of multi-scale estimation of speed. First, Simoncelli [6] proposes a bayesian scheme to compute the error distributions and then to estimate the velocity using a Kalman filter through the space of scales (not time). This approach builds a far more sophisticated error function, but it is still serial. Our work assumes that the error functions are fixed, while [6]assumes the error changes with respect to $\nabla I(\boldsymbol{x}, t)$, and this might be important in real-world scenarios. On the other hand, Chey et al. [14] propose that considering higher threshold levels for higher scales (scale-proportional thresholds) and inter-scale competition could explain human speed discrimination curves. We have presented a scheme where the response of each scale regulates the relevance of the responses of that scale. Since we handle all scales at the same time, it corresponds to a notion of threshold and competition. To our knowledge, no other work models error functions as Gaussians in the log space (this strengthen the idea that detection is not symmetrical), which seems to fit recent recordings of motion sensitivity of neurons [9].

Finally, about the time complexity of our algorithm. Lets consider the size of the image as N, the order of the optical flow algorithm as K (clearly $K(N)$) and the number of scales l. The complexity order of the serial multi-scale algorithm is $lN + lK$, and $N + lK$ for the parallel algorithm. The only difference is in the operations involved in the merge of scales. Considering the possible speed-up using p processor for the case $p = l$, then $S_p = \frac{lN+lK}{N+Kl/p}$, what can be also written as $S_p = p$. This last equation show us that the degree of parallelism (taking one level by processor) achieved by our proposed algorithm is linear.

6 Conclusions

In this work we have presented a parallel multi-scale algorithm to perform the estimation of motion using two consecutive images. This method takes bio-inspiration from human physiology and psychophysics knowledge in the sense that it achieves wide uniform relative discrimination properties by using evenly spaced logarithmic scales, and it gives results in constant time as a function of scales. With respect to the classical serial multi-scale optical flow algorithm, the error propagation among scales appears less important for our proposed algorithm in terms of relative discrimination. We now explore the idea of using more biologically plausible methods of optical flow extraction and the integration with a foveated topology.

References

1. Barron, J.L., Fleet, D.J., Beauchemin, S.S., Burkitt, T.A.: Performance of optical flow techniques. In: CVPR, vol. 92, pp. 236–242 (1994)
2. Srinivasan, M., Zhang, S.: Motion Cues in Insect Vision and Navigation, pp. 1193–1202. MIT Press, Cambridge (2003)
3. McKee, S.P., Nakayama, K.: The detection of motion in the peripheral visual field. Vision Research 24, 25–32 (1984)
4. Orban, G.A., Calenbergh, F.V., De Bruyn, B., Maes, H.: Velocity discrimination in central and peripheral visual field. J. Opt. Soc. Am. A 2(11), 1836 (1985)
5. Lucas, B.D.: Generalized image matching by the method of differences. PhD thesis, Pittsburgh, PA, USA (1985)
6. Simoncelli, E.P.: Bayesian multi-scale differential optical flow. In: Jähne, B., Haussecker, H., Geissler, P. (eds.) Handbook of Computer Vision and Applications, ch. 14, vol. 2, pp. 397–422. Academic Press, London (1999)
7. Black, M.J.: Robust incremental optical flow. PhD thesis, New Haven, USA (1992)
8. Bradski, G., Kaehler, A.: Learning OpenCV: Computer Vision with the OpenCV Library, 1st edn. O'Reilly Media, Inc., Sebastopol (2008)
9. Nover, H., Anderson, C.H., De Angelis, G.C.: A logarithmic, scale-invariant representation of speed in macaque middle temporal area accounts for speed discrimination performance. The J. of Neurosci. 25(43), 10049–10060 (2005)
10. Metha, A.B., Vingrys, A.J., Badcock, D.R.: Detection and discrimination of moving stimuli: the effects of color, luminance, and eccentricity. J. Opt. Soc. Am. A 11(6), 1697 (1994)
11. Koenderink, J.J., van Doorn, A.J., van de Grind, W.A.: Spatial and temporal parameters of motion detection in the peripheral visual field. Journal of the Optical Society of America A 2, 252–259 (1985)
12. Maunsell, J.H.R., Van Essen, D.C.: Functional properties of neurons in the middle temporal visual area (mt) of the macaque monkey: I. selectivity for stimulus direction, speed and orientation. J. Neurophysiol. 49, 1127–1147 (1985)
13. Liu, J., Newsome, W.T.: Functional Organization of Speed Tuned Neurons in Visual Area MT. J. Neurophysiol. 89(1), 246–256 (2003)
14. Chey, J., Grossberg, S., Mingolla, E.: Neural dynamics of motion processing and speed discrimination. Vision Research 38(18), 2769–2786 (1998)

Analytical Framework for Contact Time Evaluation in Delay-Tolerant Networks

Issam Mabrouki, Yezekael Hayel, and Rachid El-Azouzi

CERI/LIA, Université d'Avignon 339, chemin des Meinajaries, 84911 Avignon, France

Abstract. In the last few years, there has been an increasing concern about stochastic properties of contact-based metrics under general mobility models in delay-tolerant networks. Such a concern will provide a first step toward detailed performance analysis of various routing/forwarding algorithms and shed light on better design of network protocols under realistic mobility patterns. However, throughout the variety of research works in this topic, most interests rather focused on the inter-contact time while other contact-based metrics such as the contact time received too little interest. In this paper, we provide an analytical framework to estimate the contact time in delay-tolerant networks based on some recent key results derived from biology and statistical physics while studying spontaneous displacement of insects such as ants. In particular, we analytically derive a closed-form expression for the average value of the contact time under the random waypoint mobility model and then give an approximation for its distribution function.

Keywords: Delay-tolerant networks, Performance evaluation, Contact time, Bio-inspired networks.

1 Introduction

Delay-tolerant networks (DTN) are complex distributed systems that are composed of wireless mobile/fixed nodes, and they are typically assumed to experience frequent, long-duration partitioning, and intermittent node connection [1,2]. In such networks, communication opportunities appear opportunistic, and an end-to-end path between source and destination may break frequently or may never exist. Due to such special features, many techniques [2] have been proposed for message delivering in DTN with high probability even when there is never a fully connected path between source and destination nodes. Most of such techniques take benefit of opportunities offered by *node mobility*.

The performance of such data delivery techniques depends on the knowledge of traditional networking parameters such as node density, mobility pattern, and transmission range, to name a few. However, since DTNs differ from traditional mobile ad hoc networks in that disconnections are the norm instead of the exception, two additional critical parameters arise here [3]. The first one, which is commonly called the *inter-contact* time or sometimes referred to as the inter-meeting time, can be defined as the time duration between consecutive

E. Altman et al. (Eds.): Bionetics 2009, LNICST 39, pp. 177–184, 2010.
© Institute for Computer Sciences, Social-Informatics and Telecommunications Engineering 2010

points of time where two relay nodes come within transmission range of one another. The second one, called the *contact time* or the contact duration, can be defined as the period of time during which two nodes have the opportunity to communicate. The importance of both parameters stems from the fact that they directly impact the delay and capacity of the network, thereby helping to choose the proper design of various scheduling/forwarding algorithms for DTNs. However, throughout the variety of research works in this topic, most interests rather focused on the inter-contact time while the contact time received too little interest. This is due to empirical observations and assumptions often made in initial works that the contact time is several orders of magnitude less than the inter-contact time [4], thereby making the inter-contact time more crucial. This is our primary motivation that prompted us in this paper to focus on the contact time and try to provide an analytical framework for its estimation.

There have been various research works on the characteristics of the inter-contact time and its impact on the performance of different proposed data forwarding schemes [3,5]. Initial works typically assumed that the CCDF (complementary cumulative distribution function) of the inter-contact time decays exponentially over time under several currently used mobility models such as random waypoint model and simple random walks [6]. Although this assumption is supported by numerical simulations conducted under most existing mobility models in the literature, it is generally conjectured by authors so as to make their analysis tractable [6,7]. However, extensive empirical mobility traces later show that the CCDF of the inter-contact time follows approximately a power law over large time range with exponent less than unit [3,5].

At first glance, this finding suggested a need of new mobility models to produce the power-law property exhibited by real traces, and called for further studies to explain the outright discrepancy in the behavior of the inter-contact time. While attempting to resolve this discrepancy, many research works have recently provided credible evidence that the inter-contact time distribution has, in fact, a mixture of power-law and exponential behavior [8,9]. Specifically, using a diverse set of measured mobility traces, authors in [9] found that the CCDF of the inter-contact time follows closely a power-law decay up to a characteristic time, which confirms earlier studies, and beyond this characteristic time, the decay is rather exponential.

Knowing the inter-contact time allows one to evaluate the end-to-end delay in DTNs under ideal conditions of infinite bandwidth and buffer space. This might be a useful approximation for low traffic scenarios or low-resources data forwarding schemes. However, this is inaccurate when resources are rather limited or when the data forwarding scheme utilizes a lot of resources, which characterize several applications of DTNs. In such a scenario, the contact opportunity can be lost due to several causes such as the lack of buffer space, the limited bandwidth or simply due to MAC contention and interferences. In all these cases, even if a node comes in contact with a relay or even the destination node, it might not be able to transfer data during the contact duration. This calls to include the contact time, in addition to the inter-contact time, for a more accurate analysis of the end-to-end delay.

The remainder of this paper is structured as follows. In Section 2, we introduce basic definitions and assumptions used later to derive interesting results. In Section 3, we focus on some statistical properties of the contact time. Based on the invariance property of random walk-like motions in bounded domains encountered in many fields of science such as biology and statistical physics, we derive a closed-form expression for the average value. Under some assumptions, we give then an approximation for its probability distribution function. Finally, some conclusions are drawn in Section 4.

2   Preliminaries

We look in this section at a particular class of mobility models, namely, the random waypoint mobility model. This model is widely used for the design, study and analysis of mobile ad hoc networks. Furthermore, we introduce some useful definitions and notation and state the assumptions we will be making throughout the remaining of this paper in order to study the statistical properties of the contact time.

2.1   Random Waypoint Model

In the random waypoint mobility model [10], each node is assigned an initial location in a given area and travels at a constant speed $\mathbf{v}$ to a destination chosen uniformly in this area. The speed $\mathbf{v}$ is chosen uniformly in $[v_0, v_1]$, independently of the initial location and destination. After reaching the destination, the node may pause for a random amount of time after which a new destination and a new speed are chosen, independently of all previous destinations, speeds, and pause times. The stationary distributions of location and speed in the random waypoint mobility model differs significantly from the uniform distribution. In particular, it has been shown that the probability density function, denoted by $f_{\mathbf{v}}(s)$, for the stationary distribution of the speed without pausing is given by

$$
f_{\mathbf{v}}(s) = \begin{cases} \dfrac{1}{s \ln(v_1/v_0)} & \text{if} \qquad v_0 \le s \le v_1 \\[2ex] 0 & \text{otherwise.} \end{cases} \tag{1}
$$

2.2   Contact Criteria

There are several criteria to define a contact between two nodes. Each definition depends on the context. We restrict ourselves here to the *Boolean* and *Interference-Based* criteria [11,12] defined below.

Let $\{A_k, k \in \mathcal{T}\}$ be a set of mobile nodes following some mobility models in a common domain Ω, and simultaneously transmitting at some time instant over a certain subchannel. Let P_k be the power level chosen by node A_k, for $k \in \mathcal{T}$. Let us also denote by $A_k(t)$ the position of node A_k at time t. Consider now two arbitrary mobile nodes A_i and A_j, where $i, j \in \mathcal{T}$. Under Boolean model with communication range d, A_i and A_j are deemed to be in contact at time t if and only if the distance between them is no more than the communication range. In mathematical parlance, this can be stated as

$$\| A_i(t) - A_j(t) \| \leq d. \tag{2}$$

Note in passing that this criterion establishes a *symmetric* contact relation between nodes A_i and A_j if and only if all nodes use a common transmission range to communicate with peer nodes in the network. The major drawback of Boolean model is that it does not allow interferences to be taken into account. Although this is not generally the case for DTNs, it turns out that when the number of nodes increases, the wireless medium becomes more and more solicited. This competition for the channel may prevent successful transmissions even when the distance between A_i and A_j is no more than their minimum common transmission range. A natural way to take interferences into account is to add the sum of the interfering signals coming from simultaneous transmissions to the background noise. Then we adopt a different criterion, called the Interference-Based criterion, which can be mathematically expressed as

$$\frac{\dfrac{P_i}{\| A_i(t) - A_j(t) \|^\alpha}}{N_0 + \displaystyle\sum_{\substack{k \in \mathcal{T} \\ k \neq i}} \dfrac{P_k}{\| A_k(t) - A_j(t) \|^\alpha}} \geq \beta, \tag{3}$$

where β is a suitable is threshold, N_0 and α stand for the ambient noise and the path loss exponent respectively. Under the aforementioned contact criteria, the contact time can be formally defined as follows. Let A_i and A_j be two nodes moving according to a given mobility model. We assume that they are initially out of contact, and assume they come into contact with each other at time 0. The contact time, denoted by τ_c is defined as the time they remain in contact with each other before moving out of contact under both Boolean and Interference-Based criteria. However, for the sake of simplicity, we restrict our analysis in what follows to Boolean criterion.

3   Contact Time Analysis

In this section, we study the statistical properties of the contact time under Boolean criterion and using the random waypoint mobility model. We restrict ourselves here to the case where a relay node (or the destination node) is static all the time and thus only the source node is mobile.

3.1   Ant-Based Model Description

We consider two nodes: a mobile source node S moving at velocity $\mathbf{v}_S$ and a static relay node D (that can be also the destination node). Both nodes are assumed

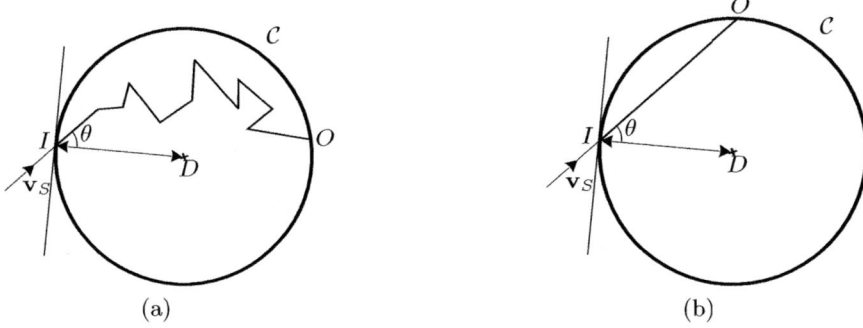

Fig. 1. (a) At point I, the source node enters the transmission region of the relay node at an angle θ to the ray ID, undergoes a random waypoint motion, and then exits at point O. (b) Source node crosses the transmission region of the relay node without changing direction.

to have the same transmission range denoted by d. We further suppose that the system is already in the steady-state, which implies that the velocity of the source node is drawn from the stationary velocity distribution characterized by the probability density function given by (1). Furthermore, it is assumed that no specific direction is favored, and therefore, that when node S enters into contact with node D, its incident direction is distributed isotropically. The contact time, denoted by τ_c, can be defined as the time elapsed from source node's entry into the radio range of relay node D until its consequent exit. In Figure 1(a), we denote by I the entrance point of the source node to the connectivity region of the relay node, namely the circle $\mathcal{C}$ centered at D with radius d. Exit point is denoted by O. Let us also denote by θ the angle that velocity $\mathbf{v}_S$ at I makes with the ray ID. We have $-\frac{\pi}{2} \leq \theta \leq \frac{\pi}{2}$, otherwise this implies the source node were already in contact with the relay node. Recalling the assumption that incident directions are distributed isotropically, θ will be uniformly distributed over $[-\frac{\pi}{2}, \frac{\pi}{2}]$. As illustrated in Figure 1(a), relay node D sees the movement of source node S as a sequence of epoch segments, where the movement direction and the velocity of the source node may change from epoch segment to another. But, the velocity remains constant in an epoch segment.

Before analyzing the contact time, we draw a parallel between the sought model and a practical animal-biology example that was of interest in many fields of science from biology [13] to statistical physics [14]. Consider ants moving on a horizontal planar surface, and assume a circle is drawn on this surface. It has been shown under isotropic incidence that the mean length of the trajectories inside the circle is independent of the random walks characteristics and is given by a very simple formula.

3.2 Mean Contact Time

Considering the random waypoint mobility model as a particular random walk and applying the above key result, a first calculus gives the mean contact time of the source node as follows $\bar{\tau}_c = \frac{\bar{L}}{\bar{\mathbf{v}}_S} = \frac{\pi d}{2\bar{\mathbf{v}}_S}$. From (1), we can readily calculate the average node speed $\bar{\mathbf{v}}_S$ and thus we obtain $\bar{\tau}_c = \frac{\pi d}{2(v_1 - v_0)} \ln(v_1/v_0)$. Many important remarks can be drawn from this key result. First, compared to previous research works [15,16], the above result is obtained without assuming that the source node has a straight line trajectory form while crossing the connectivity region of the relay node. Second, note that this above analysis can be extended somewhat to cover more general connectivity region of any geometric form around the relay node under isotropic uniform incidence. Third, in terms of the obtained mean contact time, there is no fundamental differences between the random waypoint mobility model and any other random walk-like mobility model. Therefore, the mean contact time would be the same.

3.3 Distribution of Contact Time

To avoid technical difficulty while deriving such a distribution, we consider in this paper that when the source node enters the connectivity region of the relay node, it keeps the same speed direction. As illustrated in Figure 1(b), this means that when the source node crosses circle $\mathcal{C}$, its trajectory is a chord. This assumption can be an acceptable approximation if we consider that the mean epoch distance of the random waypoint motion is higher than the diameter of $\mathcal{C}$. Furthermore, we suppose again that incident directions are distributed isotropically. It follows that the randomness of the contact time stems from the interplay of two random variables: on the one hand incidence velocity $\mathbf{v}_S$ whose probability density function is given by (1), and on the other hand angle θ that the chord IO makes with the ray ID. Let us first calculate the CDF of the length of the chord IO, denoted by L. Clearly, we have $0 \leq L \leq 2d$. Note also that L can be expressed as a function of θ and d according to $L = 2d\cos(\theta)$. Then, for all $0 \leq l \leq 2d$, we have

$$\mathbf{Pr}\{L \leq l\} = \mathbf{Pr}\{\cos(\theta) \leq \frac{l}{2d}\}.$$

Recalling that θ is uniformly distributed over $[-\frac{\pi}{2}, \frac{\pi}{2}]$, we find

$$\mathbf{Pr}\{L \leq l\} = 1 - \frac{2}{\pi}\arccos\left(\frac{l}{2d}\right).$$

Therefore, the probability density function of L can be expressed as follows

$$f_L(l) = \begin{cases} \dfrac{2}{\pi} \times \dfrac{1}{\sqrt{4d^2 - l^2}} & \text{if} \qquad 0 \leq l \leq 2d \\[2ex] 0 & \text{otherwise.} \end{cases}$$

Let us now focus on the CDF of the contact time τ_c. Note that $\tau_c = L/\mathbf{v}_S$. By conditioning on the length L of the chord IO, we find

$$\mathbf{Pr}\{\tau_c \leq t\} = \int_0^{2d} \mathbf{Pr}\{\mathbf{v}_S \geq \frac{L}{t} \mid L = l\} \times f_L(l)\, dl = \int_0^{2d} F_{\mathbf{v}}^c\left(\frac{l}{t}\right) \times f_L(l)\, dl, \quad (4)$$

where $F_{\mathbf{v}}^c(v)$ stands for the CCDF of the node source velocity, which can be readily derived from (1), so that

$$
F_{\mathbf{v}}^c(s) = \begin{cases} 1 & s \leq v_0 \\[2mm] \dfrac{\ln(v_1) - \ln(s)}{\ln(v_1) - \ln(v_0)} & v_0 \leq s \leq v_1 \\[2mm] 0 & s \geq v_1 \end{cases}
$$

Remarking that $v_0 \leq \mathbf{v} \leq v_1$ with probability one, we can calculate the integral involved in (4) by splitting it into three parts over $[0, tv_0]$, $[tv_0, tv_1]$ and $[tv_1, 2d]$ so that after elementary calculation, we obtain for all $0 < t \leq \frac{2d}{v_0}$

$$
\mathbf{Pr}\{\tau_c \leq t\} = \frac{2}{\pi} \arcsin(\frac{tv_0}{2d}) + \frac{2}{\pi \ln(v_1/v_0)} \int_{tv_0}^{tv_1} \frac{\ln(tv_1/l)}{\sqrt{4d^2 - l^2}} \, dl. \tag{5}
$$

It remains now to evaluate the integral involved in (5). Using integration by parts, we obtain

$$
\mathbf{Pr}\{\tau_c \leq t\} = \frac{2}{\pi \ln(v_1/v_0)} \int_{tv_0}^{tv_1} \frac{\arcsin(l/2d)}{l} \, dl.
$$

According to [17], we have

$$
\int \frac{\arcsin(l/2d)}{l} \, dl = \frac{l}{2d} + \frac{1}{2 \cdot 3 \cdot 3} \times \left(\frac{l}{2d}\right)^3 + \frac{1 \cdot 3}{2 \cdot 4 \cdot 5 \cdot 5} \times \left(\frac{l}{2d}\right)^5 + \frac{1 \cdot 3 \cdot 5}{2 \cdot 4 \cdot 6 \cdot 7 \cdot 7} \times \left(\frac{l}{2d}\right)^7 + \cdots
$$

Recalling that $0 \leq l \leq 2d$ and remarking that from the third order term, the coefficients involved in all higher order terms vanishes rapidly to zero, as a first approximation we retain only the linear term. Thus, we finally obtain

$$
\mathbf{Pr}\{\tau_c \leq t\} \approx \frac{(v_1 - v_0)t}{\pi d \ln(v_1/v_0)} + o(t^3) \qquad \text{for} \qquad 0 \leq t \leq \frac{2d}{v_0}.
$$

4   Conclusion

To conclude, we have addressed in this paper some statistical properties of the contact time in DTNs. Our methodology is based on a key result established in statistical physics that when a random walker enters a finite domain under isotropic uniform incidence, the mean length of its trajectories inside the domain depends only on the geometry of the system. This key result allowed us to obtain a closed-form expression for the average value of the contact time under Boolean criterion and using the traditional random waypoint mobility model. In addition, we derived an approximate formula for the probability distribution function of the contact time under the random waypoint mobility model provided that mean length of a jump is higher than the diameter of the contact area. Although this ongoing work reports a new bio-inspired methodology for the analysis of the contact time in DTNs, a number of open questions remain. Indeed, we consider to include some others promising directions in a future revision of this article.

References

1. Fall, K.: A delay-tolerant network architecture for challenged internets. In: SIG-COMM 2003: Proceedings of the 2003 conference on Applications, technologies, architectures, and protocols for computer communications, pp. 27–34 (2003)
2. Pelusi, L., Passarella, A., Conti, M.: Opportunistic networking: Data forwarding in disconnected mobile ad hoc networks. IEEE Communications Magazine 44(11), 134–141 (2006)
3. Hui, P., Chaintreau, A., Scott, J., Gass, R., Crowcroft, J., Diot, C.: Pocket switched networks and the consequences of human mobility in conference environments. In: Proceedings of ACM SIGCOMM first workshop on delay tolerant networking and related topics (2005)
4. Conan, V., Leguay, J., Friedman, T.: Characterizing pairwise inter-contact patterns in delay tolerant networks. In: Autonomics 2007: Proceedings of the 1st international conference on Autonomic computing and communication systems (2007)
5. Chaintreau, A., Hui, P., Scott, J., Gass, R., Crowcroft, J., Diot, C.: Impact of human mobility on opportunistic forwarding algorithms. IEEE Transactions on Mobile Computing 6(6), 606–620 (2007)
6. Grossglauser, M., Tse, D.N.C.: Mobility increases the capacity of ad hoc wireless networks. IEEE/ACM Trans. Netw. 10(4), 477–486 (2002)
7. Sharma, G., Mazumdar, R., Shroff, N.B.: Delay and capacity trade-offs in mobile ad hoc networks: A global perspective. In: INFOCOM (2006)
8. Cai, H., Eun, D.Y.: Crossing over the bounded domain: From exponential to power-law inter-meeting time in manet. In: MobiCom 2007: Proceedings of the 13th annual ACM international conference on Mobile computing and networking, pp. 159–170 (2007)
9. Karagiannis, T., Le Boudec, J.-Y., Vojnović, M.: Power law and exponential decay of inter contact times between mobile devices. In: MobiCom 2007: Proceedings of the 13th annual ACM international conference on Mobile computing and networking, pp. 183–194 (2007)
10. Navidi, W., Camp, T.: Stationary distributions for the random waypoint mobility model. IEEE Transactions on Mobile Computing 3(1), 99–108 (2004)
11. Gupta, P., Kumar, P.R.: The capacity of wireless networks. IEEE Transactions on Information Theory 46, 388–404 (2000)
12. Dousse, O., Thiran, P.: Connectivity vs capacity in dense ad hoc networks. In: IEEE Infocom (2004)
13. Condamin, S., Bénichou, O., Tejedor, V., Voituriez, R., Klafter, J.: First-passage times in complex scale-invariant media. Nature 450(77), 77–80 (2007)
14. Blanco, S., Fournier, R.: An invariance property of diffusive random walks. Europhysics Letters 61(2), 168–173 (2003)
15. Abdulla, M., Simon, R.: Characteristics of common mobility models for opportunistic networks. In: PM2HW2N 2007: Proceedings of the 2nd ACM workshop on Performance monitoring and measurement of heterogeneous wireless and wired networks, pp. 105–109 (2007)
16. Spyropoulos, T., Jindal, A., Psounis, K.: An analytical study of fundamental mobility properties for encounter-based protocols. Int. J. Auton. Adapt. Commun. Syst. 1(1), 4–40 (2008)
17. Harris, J.W., Stocker, H.: Handbook of mathematics and computational science, 1st edn. Springer, USA (1998)

A Formal Approach for a Self Organizing Protocol Inspired by Bacteria Colonies: Production System Application

Hakima Mellah[1], Salima Hassas[2], Habiba Drias[3], A. Raiah[4],
and A. Tiguemoumine[4]

[1] Research Center in Scientific and Technical Information, Cerist, Algeria
hmellah@mail.cerist.dz
[2] Liesp, Lyon1 university, France
Hassas@bat710.univ-lyon1.fr
[3] USTHB, Algeria
hdrias@usthb.dz
[4] Blida Univeristy, ALgeria

Abstract. Any dysfunction in production system (PS) is likely to be very expensive; so modelling by Multi Agent Systems (MAS) makes the production system (PS) possible to have aspects of robustness, reactivity and flexibility, which allow the PS control to be powerful and to react to all the risks being able to occur. In order to have a fault-tolerant PS, we propose when and how to recourse to a self organizing protocol making the MAS capable of changing its communication structure or organization, and thus reorganizing itself without any external intervention.

Keywords: Production system, self organization, MAS protocol, failure detection.

1   Introduction

Research of productivity remains the major objective of the industrial world. This objective is in permanent evolution and it requires studies and brings increasingly complex solutions in real time piloting of production systems (PS). The critical and important problem to solve is fault tolerance. The control system must provide very powerful mechanisms for fault tolerance in order to ensure the continuous operations of PS that are detection, prevention and correction, etc. A production system is a set of resources realizing a productive activity. The production is the transformation of resources (machines and materials) leading to the creation of goods or services [1]. The transformation is done by a succession of operations (tasks) that use resources (machines and operators), and modify the raw materials or components that enter into the PS in order to create outgoing finished products of this system and assigned to be consumed by customers. Changes may relate to the product form, its structure, its appearance, and so on. The transformation undergone by products, brings them an added

E. Altman et al. (Eds.): Bionetics 2009, LNICST 39, pp. 185–194, 2010.
© Institute for Computer Sciences, Social-Informatics and Telecommunications Engineering 2010

value. The resources belonging to the PS mobilized for achieving the production activity can be machines, operators, energy, information, tools, etc. The most important characteristics of a PS are flexibility, reactivity, robustness [2]. The systems theory [3] suggests decomposition of production systems into two subsystems:

(i) Information and decision subsystems that include a control portion which represents the intelligent part of the system, (ii)physical production subsystem consisting of a flows part transforming or assembling materials or entities , and a physical part representing all means necessary to carry out operations. The case study we considered is a production system (PS). We propose a MAS approach where interactions are based on a self organizing protocol that has the following features:

- It assures a decentralised control so each agent can take decisions regarding the interactions with the neighbourhood,
- It allows new communication ways when dysfunctions appear within the MAS or within the informational network. This feature gives robustness to the MAS and allows the PS to be fault tolerant.

2 Multi Agents Approaches for PS

MAS offers a new approach for modelling production systems. Instead of modelling distributed systems with programs exchanging data and commands, agent technology allows the creation of autonomous agents that communicate among themselves, negotiate sub objectives and coordinate intentions in order to achieve the objectives appropriate to the system[6]. In this context several approaches have been cited for modelling production system. In [7], a MAS platform is built for driving workshops. Different types of agents are proposed (resources agents, cell agent and product agent), they represent physical features, virtual islands or operation's sequences. A supervisor agent's role is to monitor the production process, it is assisted by a meta object agent to include new agents in the system. The principal goal of [10] does not deal with self organization. In [8] a MAS architecture in which each agent supervises a production resource is proposed. A supervisor agent is responsible for controlling the entire PS by communicating with agents supervising a production resource. The latter makes decisions about the production rules to be applied to the resources they supervise. However, the agent supervisor can intervene and tell each agent what rule to apply to achieve the overall objectives of the PS. Among the limitations of both approaches, we can quote:

- Possible saturation of the supervisor agent (too many messages from other agents of the system)can suddenly happen
- A failure within the supervisor agent causing stopping of system functioning.
- Communication cost can be very high (a message can take an important time to reach its destination, knowing that this message will go through the MAS leader agent).

In [9] a self organizing approach for manufacturing control is proposed, with our respect to the work of Bussman elaborated in this area, the dynamic routing is proposed to avoid possible congestion and jams on a machine. Agent' communication is assured by invitation and there is no indication on the system when the invitation is not successful or when an agent fails.

3   Bacteria Colony Self Organization

Traditionally the bacteria colonies push with a high level of nutritive elements [14]. A pattern in a bacteria colony is in fact an organizational structure making possible for the bacterium within its colony to communicate(bacterium-bacterium interactions), in order to fight against the adverse conditions of its environment [12]. A set of biological primitives characterizing the bacteria life has been translated to a communication model [13] based on a set of software processes considered as a MAS interaction protocol. The model takes into account only the first five primitives which seem for us to be the most basic and important within the colony, those are: the positioning, the checking, election, routing, and grouping. Each primitive will be described below, using mathematical symbols.

4   The Proposed Multi Agent system

Three types of agents are proposed: Resource Agent (RA), User Agent (UA), Interface Agent (IA):

- Resource agents for piloting a set of heterogeneous machines. They communicate with each other to achieve the production plan consisting in a set of tasks.
- User Agent or the operator. Its role is to develop a production plan of the entire production system. Its knowledge concerns resources, tasks;
- Interface Agent, acts as an interface of PS (intermediary between the RA and the operator); its knowledge are about tasks and RA.

4.1   System Functioning

The user agent develops the production plan (scheduling) which will be determined by the allocation of resources to tasks in order to achieve a product, the plan provides us with information on the task (identifier, priority, duration and precedence link). The user agent sends the plan to the interface agent that has the role of intermediary between the user agent and the various RA agents of the system. The interface agent sends to resource agents tasks to be done, each RA assures local management of its resource by integrating the piloting functions [4] in real time, namely: scheduling, execution and monitoring, each agent can perform these functions autonomously.

- **Scheduling**. RA provides the resource execution plan by respecting tasks priorities, precedence links as the execution time of tasks.
- **Execution.** RA executes the tasks affected to its resource by referring to the execution plan.
- **Monitoring.** RA assures the monitoring, detection, diagnosis, treatment and recovery in a failure case. The latter can be done by transferring tasks to other agents when a problem is detected within the system. In our approach the recovery is done by the recourse to the self organizing protocol that we propose above.

At the beginning of the system, the different RA and IA execute the main important processes of the self organizing protocol those are: positioning process, and Checking process. They allow the detection of a dysfunction within MAS. The rest of processes will be executed by RA for the treatment of failures. Figure1 shows the collaboration process between the system's agents while executing the protocol processes.

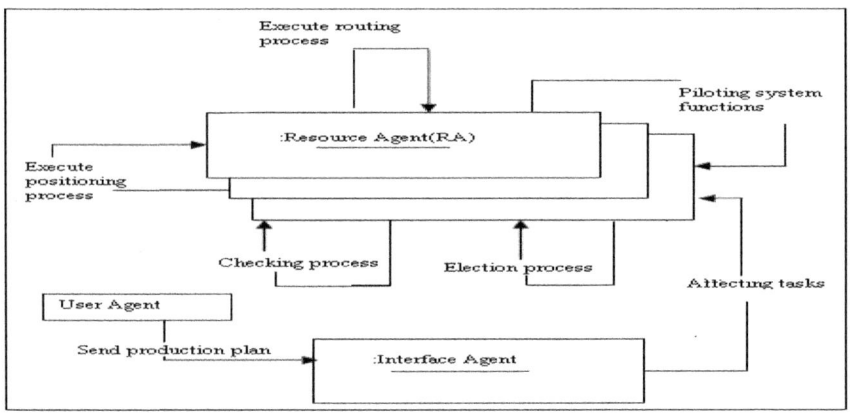

Fig. 1. Collaboration diagram of the system functioning

Fault detection and diagnosis. The fault tolerance is an important aspect characterizing a PS functioning. It is the reason for which the proposed MAS must detect and treat failures that may occur, in order to avoid the operator intervention (apart from extreme cases). The failures can be classed into two categories:

1. *Soft failure* The breakdown can be either localised when: -Communication link between two agents is destroyed. -failure is within the agent itself.
2. *Hard failure* This type of failure can be considered when malfunction of a machine happens; we call this kind of dysfunction a resource failure. The

latter is detected and diagnosed by the agent responsible of the resource below.

Failures can be detected and generally recovered through the MAS self-organising protocol without any external intervention. Detection and diagnosis of the soft failures are provided through the checking process. The checking process is carried out by RA at the system starting up.

Failures treatment. Based on the self organizing protocol, failures are:

1. *Resource Failure.* The agent responsible for this resource launches the routing process to prevent its neighbours that the agent responsible for the resource down, can not perform all its assigned tasks. So the concerned agent has to send its own address rather than sending its neighbours address. After the execution process, two cases can be found:
   -At least one RA can execute the tasks of the agent responsible for the breakdown resource (tasks are carried by the neighbourhood)
   -No agent can perform the tasks of the breakdown agent and in this case the operator has to intervene.
   If the tasks of the agent responsible for the resource down are all taken by the neighbourhood the problem is solved, otherwise, the system continues to operate without the breakdown machine pending the operator's intervention.

2. *Agent resource failure.* Two situations are quoted briefly:
   (i)RA(breakdown) is the alone agent of another RA.
   (ii)RA has several neighbours.
3. *Communication bond destroyed* When a communication link is lost, three situations can be considered:
   (i)RA isolated or the system is broken down into two subsystems.
   (ii)Isolated RA is a neighbour of an agent RA belonging to a sub system.
   (iii)The two RA belong to the same subsystem.

5   The Protocol

To make the paper self contained we present the most important processes characterizing the self organizing protocol [5] that are inspired from life within the bacteria colony. Particularly in this paper, processes are formalized in a formal manner. Each process uses a set of primitives or methods like: Leader(), groupe(), replace(), Ask(), Explore(), Life(), Rep(),.., for selecting a leader, grouping the MAS , asking for some information, exploring an agent, checking . The organization consists in N agents $A = a_1, a_2, .., a_n$, where each agent is considered as a unique node in a social network. The organization is modelled by an adjacent matrix E, where each element of E is like : $e_{ij} = 1$ if there is a communication bond between a_i and a_j , $e_{ij} = 0$ if not.

5.1 Positioning Process

Any agent in the MAS must position itself by carrying out the process of positioning. As soon as an agent integrates or leaves the group, the process of positioning is started. When an agent receives a position message: $position(a_i, role, pos_i)$, $i \in N, a_i \in A$, all agents positions and their identification are summarized in a table. $\forall j \in N, a_i \in A$, $e_{ij} = 1 \Rightarrow a_j$ is a direct neighbour of a_i and can receive a_i messages.

$\forall k \in IN, a_k \in A / e_{ik} = 1 \Rightarrow a_k$ receives $position(a_i, role, pos_i)$, If a_k receives position $(a_i, role, pos_i)$ for the first time then $Pos_{src} \leftarrow pos_i$;$pos_k \leftarrow pos_{src} + 1$;

5.2 Checking Process (Life Signal)

Local checking is an essential process. Each agent regularly diffuses a life signal to its neighbourhood. $\forall i \in N, a_i \in A, a_i$ sends $life(a_i)$ to point out that it is Kept-alive to $a_j / e_{ij} = 1$;

If $\exists a_j , a_i \in A / e_{ij} = 1$ and a_j did not receive $life(a_i)$ then a_j sends $Explore(a_i, a_j)$ to point to all the neighbourhood that it is seeking for $a_k / e_{ik} = 1$ and a_k already received $life(a_j)$. At a receipt of $Explore(a_i, a_j)$, if $a_k / e_{ik} = 1$ and a_k already received $life(a_j) \Rightarrow a_k sends Rep(response, pos_j)$ with $response = Yes/No$.

pos_j: position of the agent receiving $Explore()$ message by a_j which is searched. If $a_k / e_{kj} = 1 \Rightarrow pos_j = 1$ else $pos_j > 1$. endif.

Three situations can appear:

1. If $\exists a_k , a_k \in A / e_{ik} = 1$ and $response = Yes \Rightarrow$
   $state(a_j)$ not-in-failure
   $state(e_{ij})$ in-Failure
2. If $\exists a_k , a_k \in A / e_{ik} = 1$ and $reponse = No \Rightarrow$
   a_k is / $e_{jk} = 1$ and a_k did not receive a keep- alive signal from a_j;
   $state(a_k)$ in-failure;

3. if the received responses are:
   $(\forall k, k \neq i$ and $k \neq j$, $pos_k - pos_i \neq 1) \Rightarrow a_j$ is isolated **or**
   a_j is / $\forall k \in N, j \neq k / e_{ki} = 1$ (ie, $a_k \in$ agents group different from a_i one);
   $state(e_j) = \{$in-failure or not in-failure$\}$

Communication bond dysfunction. *(i)An isolated agent or system dissociated into two subsystems.* $State(e_{ij})$=in-failure; a_j changes its port address and sends $life(a_j)$; a_j answers by $life(a_j)$ that it agrees for the new address; in case of no possibility to change the port address, the dysfunction is certainly within the material and an alert is set off to change the link.
(ii)An isolated agent neighbour of an agent pertaining to a subsystem(Figure2).

Let set a_i the isolated agent; a_i changes its address port and sends $life(a_i)$ message to a_j, a_j replies by $life(a_j)$ to agree the new address. Only the isolated agent can change its address.

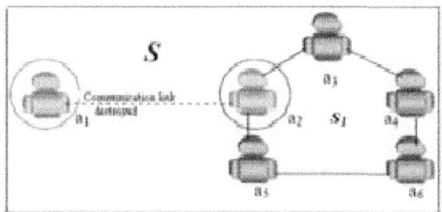

Fig. 2. An agent isolated by the destruction of its communication link

(iii)both agents pertain to the same system. If $\exists a_i$, a_j / a_i, $a_j \in A$,(a_i and a_j pertain to the same system) $\Rightarrow$ $\exists$ k $\in$ N / $a_k \in A$, $e_{ik} = 1$ and $e_{kj} = 1$, which means that it is possible to find another way to assure the communication between a_i and a_j. The latter launches the *positionning* process and the *routing* process.

Agent failure. *(i)Failed agent is the unique neighbour of another agent.* a_i is the failed agent, a_k is the unique neighbour, a_s searched agent
(a)a_k is able to execute a_i tasks, a_k launches election process. The leader selected removes a_i . a_k launches *positioning* process.
(b) a_k is unable to execute a_i tasks set, the *positioning* process is launched.
If a_k finds a_s
a_s launches *election* process
a_s launches *positioning* process
else
a_k launches *election* process. The leader removes a_i, creates a new agent, affecting it a_i position and tasks.
endif
(ii)Failed agent is a neighbour of several agents
(a) Each agent can execute a subset of tasks
$\forall$ i, k $\in$ N, a_i , $a_k \in A$, e_{ik}=1, state(e_{ik})=in-failure, $\exists$j $\in$ N / Tasks[a_j]= Tasks[a_j]+ Tasks[a_i](Each tasks subset is affected to a_k)
(b)No agent can execute a subset of tasks
$\forall$ i, k $\in$ N / a_i , $a_k \in A$, e_{ik}=1, state(e_{ik})=in-failure
$\forall$ j $\in$ N / e_{kj} =1, then a_j launches *routing process.*

5.3   Election Process

Election process is charged to select a leader agent for any decision as adding or removing an agent. It can be launched by more than one agent. As a result an agent leader is selected based on its fitness value. Once the dysfunction is located, a_i sends Leader(chefglo, fitness) to a_j, where chefglo is agent identity and fitness is the fitness value initialized to zero and incremented by 1 after a task accomplishment by an agent.

$\forall$ i $\in$ N, $a_i \in A$, Leader(chefglo, fitness) message is received and propagated in the neighbourhood until getting the highest fitness value; the leader is the agent corresponding to this value.

If $\exists i, k \in IN$/ a_i, $a_k \in A$ and fitness(a_i)= fitness(a_k), the leader is the one having the long identifier. In this case *Election process* is launched for removing or adding agents in the group and for any decision to take when necessary.

5.4 Grouping Process

In order to avoid that the election process spends too much time selecting the leader, especially when the agent number is high, the Grouping process groups agents by role, without inhibiting all the group; are inhibited only those that have the same role.

$\forall$ i $\in$ N, $a_i \in$ A, a_i sends Groupe(chefglo, fitness) message to a_i / Role(a_i)= Role(a_i).

5.5 Routing Process

The routing process is useful in two cases:

(i) When an agent (not in failure) attempts to replace the agent in failure by other/others in the neighbourhood.

The routing process allows to an agent a_i to select a_j in A where $j = 1, n \wedge j \neq i \wedge$ a_j replaces a_i;

Replace ($a_{failure}$) will be sent to all a_i / i $\in$ N, $a_i \in A$ /$i = 1, n$ and i $\neq$ failure

Task [$a_{failure}$]: tasks table of the failed agent

A variable is used to indicate if task[$a_{failure}$] is empty or not. It is initialised to false when an agent detects its neighbour in failure. While receiving Replace($a_{failure}$) with task[$a_{failure}$] updated, each neighbour verifies the content of task[$a_{failure}$]. If tasks are all under its capacities, it adds them to its tasks table, otherwise it takes those that are under its capacities and sends replace ($a_{failure}$) to the neighbours.

(ii) When an agent is lost by a destroyed communication bond, its neighbour tries to find another way to reach it, and by searching in the neighbourhood those that can serve as intermediary. *ask(a_{src}, a_{emet}, vois-info)* message is sent to its direct neighbours ie), *ask(a_{src}, a_{emet}, vois-info)* is sent to a_k / k=1, n and k $\neq$ src and k $\neq$ emet. While receiving the message for the second time by a_k, which can't be the intermediary between the sender and the source agent then a_k agent acquits the sender agent by sending acquit(nb, pos_{src}). While receiving the message for the first time and it is capable to be the intermediary, between the sender agent and the source agent, pos takes the value of its position relatively to the source that we analyze as follow:

-$pos_{src} > 1$ means that the receiver of the message is not a direct neighbour of source agent. The receiver of the message computes the set of agent that have not received the message Ask(), the non informed agent set will receive the message ask(a_{src}, a_{emet} ,vois-info).

6   Implementation Aspects

We have modelled via AUML (Agent Unified Modelling Language) the whole production system processes. The latter was tested using the NetLogo simulation tool [11]. It is a modelling programmable environment to simulate natural and social phenomena. In this paper we can not represent all the cases. Is represented below the case where an agent is in dysfunction (figure3). In the first case, by affecting the breakdown agent tasks in the neighbourhood, and without creating another one, in its localization, the production system is still working correctly until the product is achieved.

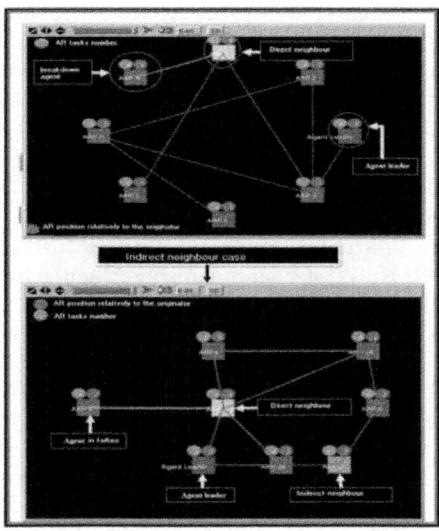

Fig. 3. An agent failure simulation

7   Conclusion

In this work a multi-agents system has been proposed for monitoring the functioning of a production system. The proposed MAS is based on a self organizing protocol[5] we have formalized using mathematical symbols. The protocol has the feature of assuring control decentralization so that each agent can take decisions to interact with its neighbours when necessary, and adapt when environment presents dysfunction. The MAS maintains its connectivity without any external intervention. Agents can perform several functions during their life as being leaders, which is not specific to a single agent. By the mean of their checking and decision they allow the emergence of new organizational structures, through agents' interactions, to cope with not desirable changes, and this is what increases the system fault tolerance and robustness.

References

1. GIARD V. Gestion de production. 2nd edn. Paris: Ed Economica (1988)
2. Draghici, G., Brinzei, N., Filpas, I.: La modélisation et la simulation en vue de la conduite des systèmes de production. Les cahiers des enseignements francophones en Roumanie (1998)
3. Habchi, G., Huget, M., Pralus, M.: D'une approche composant vers une approche agent pour un pilotage optimisé des systèmes de production. 6eConférence Francophone de MOdélisation et SIMulation, MOSIM 2006 Rabat, Maroc (2006)
4. Suarez, O.A., Foronda, J.L.A., bren, M.: Standard based framework for development of manufacturing control system. International Journal of Computer Integrated Manufacturing 11(5) (1998)
5. Mellah, H., et al.: Massop: A self organizing protocol for Multi agent systems inspired by bacteria colonies. In: CODS 2007, China, sisn.2007.07.090, July 2007, vol. 1(3), pp. 310–314 (2007)
6. Bussmann, S., Jennings, N.R., Wooldridge, M.: Multiagent Systems for Manufacturing Control: A Design Methodology. Springer Series on Agent Technology. Springer, Heidelberg (2004)
7. Roy, D.: Une architecture hiérarchisée multi-agents pour le pilotage réactif d'ateliers de production. Thèse de Doctorat: Université de Metz (1998)
8. Kouiss, K., Pierreval, H., Mebarki, N.: Using muliagent architecture in FMS for dynamic scheduling. Journal of Intelligent Manufacturing 8, 41–47 (1997)
9. Bussman, S., Schild, K., et al.: Sel-organizing Manufacturing Control: AN industrial Application of Agents technology. In: 4th Int. Conf. on MAS (2000)
10. Davidsson, P., et al.: A MAS architecture for coordination of just-in-time production and distribution. In: SAC 2002, Spain (2002)
11. Wilensky, U.: NetLogo User Manual version 3.1.1 (2006)
12. Radhika, N.: Programmable Self Assembly: constructing Global Shape using Biologically inspired local intercations and Origami mathematics. PhD Thesis, MIT (2001)
13. Mellah, H., et al.: A communication model of distributed information sources bacteria colonies inspired. In: 10th IEEE International Conference on Intelligent Engineering Systems, INES 2006, London,UK (2006)
14. Jacob, E.B., et al.: Modelling branching and chiral colonial Patterning of Lubricating Bacteria. In: Proceedings of IMA workshop: Pattern formation and Morphogenesis (1998)

Delay Tolerant Networks in Partially Overlapped Networks: A Non-cooperative Game Approach⋆

Rachid El-Azouzi, Habib B.A. Sidi, Julio Rojas-Mora, and Amar Prakash Azad

LIA, Université d'Avignon 339, chemin des Meinajaries, 84911 Avignon, France

Abstract. Epidemic forwarding protocol in Delay Tolerant Networks maximizes successful data delivery probability but at the same time incurs high costs in terms of redundancy of packet copies in the system and energy consumption. Two-hop routing on the other hand minimizes the packet flooding and the energy costs but degrades the delivery probability. This paper presents a framework to achieve a tradeoff between the successful data delivery probability and the energy costs. Each mobile has to decide which routing protocol it wants to use for packet delivering. In such a problem, we consider a non-cooperative game theory approach. We explore the scenario where the source and the destination mobiles are enclosed in two different regions, which are partially overlapped. We study the impact of the proportion of the surface covered by both regions on the Nash equilibrium and price of anarchy. We also design a fully distributed algorithm that can be employed for convergence to the Nash equilibrium. This algorithm does not require any knowledge of some parameter of the system as the number of mobiles or the rate of contacts between mobiles.

1 Introduction

Delay tolerant mobile ad-hoc networks have gained attention in recent research. Instantaneous connectivity is not needed any more and messages can arrive at their destination thanks to the mobility of some subset of nodes that carry copies of the message. A naive approach in forwarding a message to the destination consists in the use of an epidemic routing strategy, in which any mobile that has the message keeps on relaying it to any other mobile that arrives within its transmission range and which does not still have the message. This would minimize the delivery probability at a cost of inefficient use of network resources in terms of energy used for transmission. The need for a more efficient use of network resources has motivated the use of more economic packet forwarding strategies such as the two-hop routing protocols, in which the source transmits copies of its message to all mobiles it encounters, but these relay the message only if they come in contact with the destination. The performance of the two-hop forwarding protocol along with the effect of the timers have been evaluated in [1]. In this paper we consider an alternative approach that offer a way of studying the successful delivery probability and energy consumption. This paper aims to provide a scheme which maximizes the expected delivery rate while satisfying a certain constant on the number of forwardings per message. To do this, we assume that each mobile may decide which

---

⋆ This work has been partially supported by Bionets project.

E. Altman et al. (Eds.): Bionetics 2009, LNICST 39, pp. 195–202, 2010.

routing protocol it wants to use for delivering packets. We restrict the case that only two routing protocols are available to mobiles: epidemic routing and two-hops. This scheme allows us to exploit the trade-off between delivery delay and resource consumption. The higher number of users use epidemic (resp. two hops) routing , the higher (resp. lower) probability of success and the higher (resp. lower) consumption of resource.

In our study we assume that each mobile like to find the routing protocol that maximizes his utility function. But, as this utility depends on the action of the other mobiles, the system can be described as a non-cooperative game. We show that this game has at least one Nash equilibrium, and we designed a distributed algorithm to reach it. This algorithm is implemented at each node, allowing the system to reach the Nash equilibrium in a completely distributed way. Since the estimation of some parameters of the system, is very difficult in DTN, due to the lack of persistent connectivity, the proposed algorithm also allows the nodes to converge to the Nash equilibrium without any information.

Delay Tolerant Networks (DTNs) have recently attracted attention of the research community. Delay Tolerant Networks (DTNs) are sparse and/or highly mobile wireless ad hoc networks where no continuous connectivity guarantee can be assumed [2, 3]. There are several results of real experiments on DTNs [6, 11, 13]. In [10], the authors studied the optimal static and dynamic control problems using a fluid model that represents the mean field limit as the number of mobiles becomes very large. In [9], the optimal dynamic control problem was solved in a discrete time setting. The optimality of a threshold type policy, already established in [8] for the fluid limit framework, was shown to hold in [9] for the actual discrete control problem. A game problem between two groups of DTN networks was further studied in [9].

2   The Model

We consider two overlapping network regions, where source and destination nodes are each in distinct regions. By network region we mean a region with moving nodes that can establish a connection between them. We assume that nodes have random way-point mobility (see [7]) which is confined to the region it is associated . In context of DTN the transportation of data relies mainly on mobility, so the overlapping region plays an important role. Overlapping regions are the only place where nodes can exchange data from one region to another. Consider that network region S_1 contains a source S, and N_1 mobile nodes, and that network region S_2 contains the destination node d and N_2 mobile nodes. Since source and destinations are in different regions, data can be transported from source to destination by mobile nodes only through the overlapping region $\hat{S}$. Let us parameterize the overlapped(normalized) region, denoted by $\tilde{S} = \hat{S}/\max\{S_1, S_2\}$. Notice that the overlapping region $\tilde{S}$, when parameterized reduces to (assume $S_1 = S_2$ for simplicity) the following special cases : "Unified network", i.e., when $\tilde{S} = S_1 = S_2$, and "Overlapped network" when $0 < \tilde{S} < 1$.

We assume that each mobile node is equipped with some form of proximity wireless communications device. The network regions are assumed to be sparse, so that, at any time instant, nodes are isolated with high probability. Communication opportunities arise whenever, due to mobility patterns, two nodes get within mutual communication

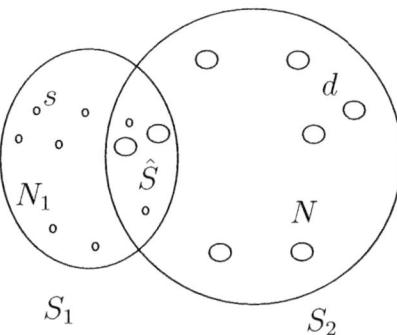

Fig. 1. Overlapped Network Region $\hat{S}$

range. We refer to such events as "contacts". The time between subsequent contacts of any pair of nodes is assumed to follow an exponential distribution. The validity of this model for synthetic mobility models (including, e.g., Random Walk, Random Direction, Random Waypoint) has been discussed in [1]. In [7], the authors derived the following estimation of the pairwise meeting rate λ :

$$\lambda = \frac{2wRE[V^*]}{S}, \tag{1}$$

where w is a constant specific to the mobility model, $E[V^*]$ is the average relative speed between two nodes and R is the range. Let λ_1 (resp. λ_2) be the rate of meeting of any pair of nodes in region S_1 (resp. S_2). Let λ_S denote the rate of meeting between the source and a node in region S_2. From (1), we have

$$\lambda_1 = \frac{2wRE[V_1^*]}{S_1}, \quad \lambda_2 = \frac{2wRE[V_2^*]}{S_2} \text{ and } \lambda_s = \frac{2wRE[V_s^*]}{S_1}.$$

Similarly, the rate of meeting between a node (resp. source) in S_1 and a node in S_2 is given by $\lambda_{12} = \frac{2wRE[V_{12}^*]}{\hat{S}}$, $\lambda_{s_2} = \frac{2wRE[V_{s_2}^*]}{\hat{S}}$,

where V_{s_2} is the average relative speed between source and a node in region S_2. There can be multiple source-destination pairs, but we assume that at a given time there is a single message, eventually with many copies, spreading in the network. For simplicity we consider the message originated at time t = 0. We also assume that the message that is transmitted is relevant only during some time τ. The message contains a time stamp reporting its generation time, so that it can be deleted at all nodes when it becomes irrelevant.

A mobile terminal is assumed to have a message to send to a destination node. We consider in this paper two types of routing in DTN networks: epidemic routing and two-hop routing. In this paper we study the competition between individual mobiles in a game theoretical setting. Each mobile can decide whether to use epidemic or two-hop routing, depending on which strategy maximizes his utility function. We assume that the source node S stays in region S_1 while the destination node d stays in region S_2. Naturally, the nodes in S_1 needs to forward the packet to the nodes in S_2. Hence, the nodes in S_1 are of "Epidemic" type only, while nodes in S_2 may be of either type.

Consider that there are N_1 mobiles among the total N_{tot_1} in region S_1 which participate in forwarding the packet using epidemic routing. We assume that N mobiles among N_{tot} in region S_2 can choose between epidemic and two-hop routing. Let N_e^0 (resp. N_t^0) be the number of mobiles that always use epidemic (resp. two-hop) routing. Then, we have:

$$N_{tot} = N + N_e^0 + N_t^0$$

The source in region S_1 has a packet generated at time 0 that wishes to send to the destination d in region S_2. In region S_2, let N_e (resp. N_t) be the number of users that use epidemic routing (resp. two-hop routing). Let $X_e(t)$ (resp. $X_t(t)$) be the number of mobile nodes (excluding the destination and source) that use epidemic routing (resp. two-hop) and have at time t a copy of the packet. Denote by $D_i(\tau)$ the probability of a successful delivery of the packet by time τ. Then, given the process X_i (for which a fluid approximation will be used), we have the probability of successful delivery of packet as:

$$P_{succ}(\tau) = 1 - e^{\left(-\lambda_d \int_0^\tau (X_e(t)+X_t(t))dt\right)} \qquad (2)$$

where λ_d denotes the inter-meeting rate between the destination and a node in S_2. Consider that on successful delivery of the packet is rewarded with $\bar{\alpha}$ which is shared among all the participating nodes. Let the reward is shared among the two region as α_{S_1} for region S_1 and α for S_2, where $\bar{\alpha} = \alpha_{S_1} + \alpha$. In region S_1 there are only epidemic type user, the reward is shared equally among $X_1(\tau)$ users. While in region S_2, the reward α is further shared as α_e (resp. $\alpha_t = \alpha - \alpha_e$) among the mobiles that have at time τ a copy of the message and use epidemic (resp. two-hop) routing. Hence, the utility U_e (resp. U_t) for a player using epidemic (resp. two-hop) routing is given by

$$U_e(N_e) = \left(\frac{\alpha_e P_{succ}(\tau)}{X_e(\tau)} - \beta\tau\right)\mathbb{P}_1\left(\text{resp. } U_t(N_e) = \left(\frac{\alpha_t P_{succ}(\tau)}{X_t(\tau)} - \gamma\tau\right)\mathbb{P}_1\right) \qquad (3)$$

where β and γ are the energy cost ,and $\mathbb{P}_1(t) = 1 - e^{-\int_0^t (\lambda_{s_2}+\lambda_{12}X_1(s)+\lambda_2 X_e(s)ds)}$ which denotes that the probability of receiving a packet by time t.

2.1 Fluid Approximation

We consider the following standard fluid approximation (based on mean field analysis)

$$\frac{dX_1(t)}{dt} = (\lambda_s + \lambda_1 X_1(t) + X_e(t)\lambda_{21})(N_1 - X_1(t)), \qquad (4)$$

$$\frac{dX_e(t)}{dt} = (\lambda_{s_2} + \lambda_{12}X_1(t) + X_e(t)\lambda_2)(N_e - X_e(t)), \qquad (5)$$

$$\frac{dX_t(t)}{dt} = (\lambda_{s_2} + \lambda_{12}X_1(t) + X_e(t)\lambda_2)(N_t - X_t(t)). \qquad (6)$$

The essage is spread directionally, which means that nodes from region S_1 can forward the packet to nodes in S_2, while the reverse is not allowed,so $\lambda_{21} = 0$. On solving the ODE's given in eq. (4)-(6) using the suitable initial conditions, we obtain

$$X_1(t) = \frac{\lambda_s N_1 \left(1 - \exp\left(-t\left(\lambda_s + \lambda_1 N_1\right)\right)\right)}{\lambda_s + \lambda_1 N_1 \exp\left(-t\left(\lambda_s + \lambda_1 N_1\right)\right)}, \tag{7}$$

$$X_e(t) = \frac{N_e\left[\psi(t)\left(1 - N_e \int_0^t \frac{\lambda_2}{\psi(u)} du\right) - 1\right]}{\psi(t)\left(1 - N_e \int_0^t \frac{\lambda_2}{\psi(u)} du\right)}, \tag{8}$$

$$X_t(t) = N_t\left(1 - \exp\left[-\lambda_{12}\int_0^t X_1(u)du + \lambda_2\int_0^t X_e(u)du + t\lambda_{s_2}\right]\right). \tag{9}$$

where $\psi(t) = \exp\left(\int_0^t \left(\lambda_{s_2} + \lambda_{12}X_1(u) + \lambda_2 N_e\right) du\right)$.

3 The DTN Game

As explained before, there is but   a single choice for the nodes in region S_1, i.e., to participate or not in epidemic forwarding. However in region S_2, a node can choose between participating or not, and, if so, it can choose between epidemic forwarding or two hop forwarding to deliver the packet to destination. Every mobile would like to find the strategy that maximizes his individual utility. But, as his utility depends on the actions of the other mobiles, the system can be described as a non-cooperative game. As the game is symmetric, a Nash equilibrium (NE) N_e^* is given by the two conditions:

$$U_e(N_e^*) \geq U_s(N_e^* - 1) \text{ and } U_t(N_e^*) \geq U_e(N_e^* + 1)$$

The previous definition means that no user using epidemic routing (resp. two-hop routing), has an incentive to use two-hop routing (resp. epidemic routing). The existence of the Nash equilibrium is guaranteed by [12].

4 Stochastic Approximation for Nash Equilibrium

In this section we introduce a distributed method to achieve the Nash equilibrium in the case where some parameters (i.e., N, λ and λ_s) are unknown. We show that simple iterative algorithms may be implemented at each node, allowing them to discover the Nash equilibrium in spite of the lack of information on such parameters. Note that the estimation of N, λ and λ_s, is very difficult in DTN because of the lack of persistent connectivity. This distributed algorithm proposed in [5] was proved, for a fixed number of players, that if it converges, it will always do to a Nash equilibrium. In order to increase the speed of convergence, each user decides to stop his update mechanism after reaching a given threshold [4]. It is not a global convergence criteria, as we can find in centralized algorithms, but an individual convergence criteria that let each user stop calculations. The algorithm is based on a reinforcement of mixed strategies and players are synchronized in such a way that the decision of all players (playing pure strategy) induce the utility perceived for each one.

The algorithm works in rounds. Each round corresponds to the delivery of a message by the source. Let $N_e(t)$ be the number of players that use epidemic routing at round t.

At each round t, each user i chooses epidemic routing over the set $C = \{e, t\}$ of strategies, with probability p_t (and chooses the two-hop routing with probability $1 - p_t$). The utility perceived by user i at round t depends on his action and on the actions of the other mobiles. This utility u_t^i is expressed as follows:

$$u_t^i = \mathbb{1}_{\{c_t = e\}} \cdot U_e(N_e(t)) + \mathbb{1}_{\{c_t = t\}} \cdot U_t(N_e(t)) \tag{10}$$

Then, each player updates his probability according to the following rule (see Algorithm 1):

$$p_t^i = p_{t-1}^i + b \cdot \left(\mathbb{1}_{\{c_t = e\}} - p_{t-1}^i \right) \cdot u_t^i, \tag{11}$$

Figure 3.b shows the evolution of the probabilities and the convergence to Nash Equilibrium for a set of 10 players, using a treshold of convergence at $\epsilon = 10^{-6}$.

5 Global Optimum Repartition and Nash Equilibrium

In this section, we are interested in the network efficiency as the maximization of the global optimum of the system. We want to optimize the overall network energy-efficiency with respect to the aforementioned degrees of freedom. For this purpose, we consider the optimal social welfare, which is well known in game theoretic studies, and compare it with the performance achived at Nash Equilibrium.

The following simulations allow us to see the range of values for different parameters which minimizes the gap in total utility between the Nash equilibrium and the global optimum. For different rates of λ_s and different values of the reward on epidemic routing α_e, we compute the price of anarchy, using the total utility at the global optimum repartition and at Nash Equilibrium.

The social welfare of the network is measured by the total utility of the system expressed by

$$W_s = X_e(\tau)U_e(N_e) + X_t(\tau)U_t(N_e) \tag{12}$$

and the price of anarchy is measured as follows:

$$PoA = (W_s^{Opt} - W_s^{NE})/W_s^{Opt} \tag{13}$$

where W_s^{Opt} (resp. W_s^{NE}) is the social welfare at the global optimum (resp. at the Nash Equilibrium.)

Through the different simulations for several set of values for the main parameters of our DTN network, we observe the network stability and efficiency. In figure 2 we plot the evolution of the number of users infected using either two hops or epidemic routing. As we can notice, the rate of infection of users using epidemic routing increases with the inter-meeting rate in the second region before reaching a stability point, that is mainly influenced by the relevant time of packet delivery which increases the probability of success and makes the infection rate independent on λ_2. This rate is always bigger with the surface of overlapping and the reward on using epidemic routing. We observe the same behavior for the infection rate of users using two-hop routing, except that the infection rate become smaller with the reward on using epidemic routing. Figure 3.a present on the other hand the price of anarchy (PoA) at Nash Equilibrium. For small

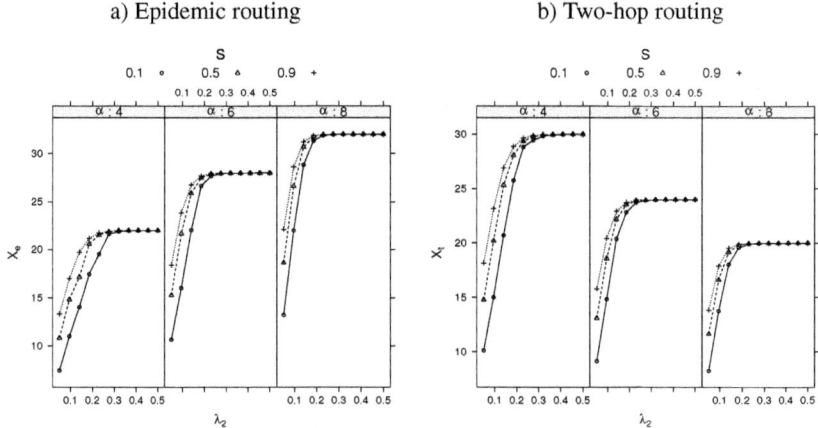

Fig. 2. Infected users using epidemic or two-hop routing

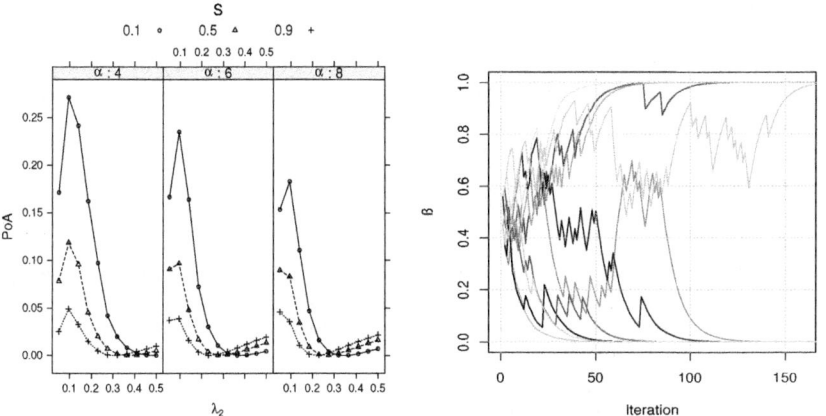

Fig. 3. a) Price of anarchy depending on λ_2 b) Convergence to Nash Equilibrium

values of the inter-meeting rate λ_2 in the second region, the PoA takes it highest values and is almost independent on $\tilde{S}$. The optimality of the Nash equilibrium (obtained when the PoA is near or equal to zero) is achived for small values of λ_2 by increasing α_e or $\tilde{S}$.

6   Conclusion

This paper presents a framework to analyse the tradeoff between the successful data delivery probability and energy costs. We formulate the problem as a non-cooperative game in which each mobile has to decide which routing protocol it wants to use for packet delivering: Epidemic routing or Two-hop routing. We explore the scenario where the source and the destination mobiles are enclosed in two different regions, which are partially overlapped. We showed the impact of overlapping area on price of anarchy and Nash equilibrium. To complete this contribution, we plan to analyze the system when there are new arrivals to the area of interaction and mobiles within this area will be active for a limited period of time. This configuration makes the system dynamic in the number of mobiles, a more realistic approach to a DTN case.

References

1. Al-Hanbali, A., Nain, P., Altman, E.: Performance of Ad Hoc Networks with Two-Hop Relay Routing and Limited Packet Lifetime. In: First International Conference on Performance Evaluation Methodologies and Tools (Valuetools), Pisa (2006)
2. Zhao, W., Ammar, M., Zegura, E.: Controlling the Mobility of Multiple Data Transport Ferries in a Delay-Tolerant Network. In: Proceedings of IEEE INFOCOM 2005, Miami, Florida (2005)
3. Jain, S., Fall, K., Patra, R.: Routing in a Delay Tolerant Networking. In: Proceedings of SIGCOMM 2004 (2004)
4. Coucheney, P., Touati, C., Gaujal, B.: Fair and efficient user-network association algorithm for multi-technology wireless networks. In: Proceedings of INFOCOM 2009 Mini Conference (2009)
5. Sastry, P.S., Phansalkar, V.V., Thathachar, M.A.L.: Decentralized learning of nash equilibria in multi-person stochastic games with incomplete information. IEEE Transactions on Systems, Man and Cybernetics 24(5), 769–777 (1994)
6. Burgess, J., Gallagher, B., Jensen, D., Levine, B.N.: MaxProp: Routing for Vehicle-Based Disruption-Tolerant Networks. In: Proceedings of IEEE Infocom (2006)
7. Groenevelt, R., Nain, P.: Message delay in manets. In: ACM SIGMETRICS, Banff, Canada, June 2005, pp. 412–413 (2005)
8. Altman, E., Basar, T., De Pellegrini, F.: Optimal monotone for- warding policies in delay tolerant mobile Ad-Hoc networks. In: Inter-Perf 2008: Workshop on Interdisciplinary Systems Approach in Performance Evaluation and Design of Computer and Communication Systems, Athens, Greece (October 2008)
9. altman, E., Neglia, G., De Pellegrini, F., Miorandi, D.: Decentralized Stochastic Control of Delay Tolerant Networks. In: IEEE Infocom, Rio de Janeiro, Brazil, April 19-25 (2009)
10. Zhang, X., Neglia, G., Kurose, J., Towsley, D.: Performance modeling of epidemic routing. Comput. Netw. 51(10), 2867–2891 (2007)
11. Demmer, M., Brewer, E., Fall, K., Jain, S., Ho, M., Patra, R.: Implementing Delay Tolerant Networking. Technical report, IRB-TR-04-020, Intel Corporation (December 2004)
12. Rosenthal, R.W.: A class of games possessing pure-strategy Nash equilibria. Internat. J. Game Theory v2 i1, 65–67
13. Greifenberg, J., Kutscher, D.: RDTN: An Agile DTN Research Platform and Bundle Protocol Agent. In: van den Berg, H., Heijenk, G., Osipov, E., Staehle, D. (eds.) WWIC 2009. LNCS, vol. 5546, pp. 97–108. Springer, Heidelberg (2009)

Author Index

GPSR Compliance

The European Union's (EU) General Product Safety Regulation (GPSR) is a set of rules that requires consumer products to be safe and our obligations to ensure this.

If you have any concerns about our products, you can contact us on ProductSafety@springernature.com

In case Publisher is established outside the EU, the EU authorized representative is:

Springer Nature Customer Service Center GmbH
Europaplatz 3
69115 Heidelberg, Germany

Batch number: 09490872

Printed by Printforce, the Netherlands